Chevrolet G-506

1½ Ton, 4x4 Development, Production and Variants in WW2

by David Doyle

PORTRAYAL
PRESS

Chevrolet G-506, 1½ Ton, 4x4 Development, Production and Variants in WW2

Published by
Portrayal Press LLC
PO Box 944
Branchville, New Jersey 07826

ISBN: 978-0-938242-06-2 (Hardbound)
ISBN: 978-0-938242-28-4 (Softbound)

Printed in the USA
Signature Book Printing
www.sbpbooks.com

PORTRAYAL
PRESS

www.portrayalpress.com

Table of Contents

Acknowledgements

Creating this book took seven years, and was far from a one-person job. Rather, compiling this required a great deal of help from a number of colleagues and friends. Browsing even briefly through the book reveals that it contains dozens of never seen before period photos of the 1 ½-ton Chevrolet. A closer examination will reveal that the text also provides the reader with previously unpublished data.

While this new material comes from a variety of sources, much of it comes straight from the source – General Motors. Access to this material was arranged by Keith Nattress, to who I am deeply indebted, and whose staff, including Larry Kinsel, and Christo Dantini provided immeasurable help in bringing this project to life.

Also instrumental in creating this book has been the help of Tom Kailbourn, Jim Gilmore, Dana Bell, Scott Taylor, Pat Ware, Jim Davis, Keith Buckley, Taigh Ramey, Eric Reinert, Ted Heinbuch, and the late Richard Adelman. Their generous and skillful assistance adds immensely to the quality of this volume.

Beyond General Motors, several other institutions have contributed to this project, including the National Archives and Records Administration, TACOM LCMC History Office, the Military History Institute, US Army Communications-Electronics Command, the US Army Transportation Museum, Quartermaster Museum and Ordnance Museum, and the National Museum of the United States Air Force.

In addition to such wonderful friends and colleagues, the Lord has blessed me with a wonderful wife, Denise, who has tirelessly scanned thousands of photos and documents for this and numerous other books. Beyond that, she is an ongoing source of support and inspiration.

Preface

Despite production of over 150,000 examples during WW2, little documentation has survived to reveal the tale of the design, production and evolution of the 1½-ton Chevrolet 4x4 truck.

Perhaps this should not be surprising given the passage of seventy + years. After all, how much of your current office paperwork would you expect to be locked away, untouched, 70 years from today?

While we now consider these vehicles historic, in the 1940s both General Motors and the U.S. Government considered them purely utilitarian. Under wartime pressure, vast amounts of paperwork were generated to coordinate production at Chevrolet plants as well as countless subcontractors. Hordes of civilian and military personnel took care of dispatching these trucks to supply armys throughout the world and documented each such movement. A mass of workers tended to the flow of all of this paperwork, which must have filled many, many filing cabinets.

On VJ-Day the number of employees available to shepherd this paperwork was reduced significantly. Not only that, as a war-weary nation, its economy flush for the first time since prior to the Depression, sought to fill their garages with new cars, documentation of war production became increasingly simply "stuff in the way" – a mindset that increased with each passing year. This is to a large extent inevitable as the people involved were simply doing their job. Details that are important to enthusiasts today were merely routine to the men at the time. Couple this sentiment with consolidation of manufacturing plants and divisions, and the expense that would have come with packing, moving, sorting and storing these records, and it is remarkable that ANY documentation survived.

Over many weeks at the General Motors archive researching the GMC CCKW and DUKW for my other titles on those vehicles, I came across remnants of Chevrolet production records. Whereas GMC plant manager Russell Crist had carefully created five copies of a four volume contemporaneous record of that GM Division's war efforts – and remarkably some of those copies survived - it seems that there was no such champion to similarly document Chevrolet's endeavors.

In the absence of a "Crist-like" contemporaneous history or the actual production records, we are left to interpret shards of corporate documentation and scattered government records – the latter made doubly difficult due to changes in U.S. Army vehicle procurement and maintenance channels during the war. The 1 ½-ton Chevrolet was conceived under the auspices of the Quartermaster Corps, yet the bulk of the production took place under the authority of the Ordnance Department. The army records concerning vehicle development and production immediately prior to responsibility for wheeled vehicles being transferred from Quartermaster to Ordnance, which occurred on 25 July 1942, are sparse. This leads this writer to believe that they were either destroyed, or discarded, or are misfiled en masse at a government facility. After years of looking, I have turned up very little. One day, perhaps.

With that being said – what we were able locate in General Motors and government archives, and are now presenting in this book, reveals a great deal and substantially more than any prior volume on the subject. In many cases, the record is clear – and is stated as fact. In a few cases, I am forced to speculate, and I endeavor in these pages to clearly differentiate my speculation versus the historic record.

I will add a couple of final points to aid in putting all of this in perspective. First of all, there is a popular myth that no civilian vehicles were produced during the war years. This is exactly that – a myth. Civilian vehicle production by Chevrolet resumed in 1944. In that year the government permitted Chevrolet to build 33,122 non-military trucks. A related myth propagated by some holds that G-506 production carried over into 1946. Once again, the record is clear, G-506 production ended in 1945.

The second major myth has it that American industry, especially the automotive industry, operated at capacity during WWII. Thanks to patriotic wartime propaganda, our mind's eye envisions plants working around the clock, smoke spewing from stacks while a veritable army of laborers works at break-neck speed to produce needed implements of war. However, a wartime Chevrolet document sets the record straight, saying "The Army's allocation of Military Truck production to Chevrolet did not justify the operation of all assembly plants. Total Government monthly requirements were less than a tenth of Chevrolet's normal civilian output before the war." With suppliers able to satisfy war demands and the end of hostilities in sight, Chevrolet and the other manufacturers began to focus on growing consumer demand for vehicles.

Finally, this is not a "restoration guide" per se. We don't say that one should use this wiper blade after that serial number, or paint the oil filter a particular shade of a particular color. Such statements require documentation at an extreme level, which as mentioned previously evidently no longer exist, or extensive speculation by the author. Rather, through the lenses of cameras 75 years ago, we present evidence of what was done at the factory and in the field, and leave any speculation to the reader.

David Doyle
Memphis, Tennessee

Introduction

While on 25 July 1942, responsibility for procurement of all types of army ground vehicles was shifted to the Ordnance Department, such was not the case during the 1930s. At that time, the Army Ordnance Department had charge of the combat vehicles, while responsibility for the general purpose vehicles fell to the Quartermaster Corps. General purpose vehicles were further subdivided into two groups, administrative vehicles and tactical vehicles. Administrative vehicles tended to be commercially available vehicles, the type one would see on a dealer's showroom, or routinely on a factory assembly line. Tactical vehicles, owing to their intended field use, as a rule were equipped with all-wheel drive, a feature not commonly seen in the civilian marketplace at that time.

From the mid-1920s until well into the Cold War era, US Army weapons and material were cataloged through a system known as the Standard Nomenclature List (SNL). Within the SNL were various groupings – Group A was automatic weapons, small mortars, carts and light artillery. Group D was heavy field artillery, and most important for the purpose of this book, Group G was tank and automotive material. A numeric suffix was added to the SNL group to define various families of vehicles.

As a rule, the various series of vehicles within each family or numeric suffix have similar power trains and chassis, but the bodies vary – sometimes widely. However, as with so many 'rules', there are a few exceptions to this. Of course the Chevrolet 1 ½ ton 4x4 is one of these exceptions as the various models were designated either G-85 or G-506.

The Chevrolet 1 ½ ton bomb service truck – which from the outset was Ordnance Corps responsibility – was assigned SNL G-85. In one of the previously mentioned exceptions, that same SNL number was also assigned to bomb service trucks built by Diamond T, Ford, and Yellow Truck and Coach, despite the vehicles being vastly different from each other.

Later, as Ordnance became responsible for all vehicle procurement, a new series of G-numbers began to be used. The remaining 1½-ton 4x4 Chevrolets were placed in this category, which is SNL G-506. That term will be used extensively in this volume. The 1½-ton Chevrolet was classified "standard" by Ordnance Committee Minutes item 19107 on 23 October 1942.

Coincidentally, the internal Chevrolet model numbers for these trucks also began with a G – as in G-4112, G-7117, etc. The coincidence of both the Standard Nomenclature List number and Chevrolet model number beginning with the letter G has caused considerable confusion through the years.

Finally, into this alphabet soup should be added the War Department Catalogs, three types of which are particularly relevant to a discussion of these vehicles. They are:

- WD CAT. ORD 7 Organizational Maintenance Allowances.
- WD CAT. ORD 8 Field and Depot Maintenance Allowances.
- WD CAT. ORD 9 List of all service parts.

The three books named above are the parts books relative to vehicles, with the ORD 9 being the most useful volume today. While in military service, the Ord 7 and Ord 8 were useful for maintaining stock levels, but are of limited use today.

Thus, for the 1-1/2 ton 4x4 Chevrolet, the list of all service parts is the ORD 9, SNL G-85 (Vol. IV) and G-506. This is a single book of over 500 pages. As an aside, the Volume IV notation is a result of different volumes pertaining to bomb service trucks of different manufacturers.

During the 1930s, military procurement of motor vehicles of any type was sparse. The Great Depression, which gripped the nation, gripped the Army as well. The Army's motor vehicle fleet, which was composed at that time of maintenance-intensive WWI-era trucks of a variety of non-standard designs, badly needed updating. However, funding for such an upgrade was simply not available.

The first breakthrough in this debacle surprisingly came through the Public Works Administration, which

At the Chevrolet plant in Kansas City, workmen are making adjustments to a very early Chevrolet 1 1/2-ton 4x4 Cargo Truck: note the lack of a battery box to the rear of the cab. The man in front of the bumper is touching up paint on the fastenings for the brush guard. Registration numbers are present on the hoods but are not fully legible. The Kansas City plant shipped a total of 3,164 Chevrolet 1 1/2-ton 4x4 vehicles, all between the years 1940 and 1942. General Motors LLC

provided $10 million in funding for military vehicles as part of the New Deal. Because of the funding source, the procurement of these vehicles was governed by the National Recovery Act, which meant that price would be the deciding factor, rather than military characteristics. As part of the 1935 fiscal year appropriations bill, Congress stipulated that the Army must rid itself of all motor vehicles procured prior to 1 January 1920. This amounted to about 5,000 vehicles, including the backbone of the army's truck fleet, the Standard B, many of which had recently been rebuilt and modernized.

While this procurement initiated through the New Deal did provide the Army with 6,000 new vehicles, there was little standardization. A variety of vehicles of various weight classes from many manufacturers crowded the ranks, recreating the maintenance logistics nightmares of World War I, much to the dismay of the Quartermaster Corps, which felt hamstrung by the edict of the office of the Comptroller General, whose procurement policies had created this situation.

While in the 1920s and early 1930s, the Quartermaster Corps had wanted to manufacture in house standardized vehicles in a series of five classes to meet all Army requirements, the Comptroller General of the United States, J. R. McCarl, believed that such production would be illegal, believing instead that the Army was bound to buy vehicles from industry, on a low bid basis. This would inherently mean that there would be little standardization of the vehicle, as each contract for a given type of vehicle could be won by a different firm each time a bid was solicited. The Quartermaster Corps vehemently disagreed with the Comptroller's position, citing the National Defense Act. However, in September 1933 then-new Army Chief of Staff Douglas MacArthur issued General Order Number Nine, requiring the Army to purchase "complete vehicles from the automotive industry...."

By 1936 the General Staff had in place the Protective Mobilization Plan (PMP). An integral part of that plan was the creation of the Initial Protection Force (IPF), a 400,000-man force that within one month of mobilization would be combat-ready. Similar to today's rapid deployment force, the IPF could, in the event of war, work with local assets to fight delaying actions while the rest of the nation mobilized.

The Protective Mobilization Plan required 190,000 trucks. In 1938, speaking to the meeting of the Society of Automotive Engineers concerning the Army's 10,000-vehicle fleet prior to implementation of the PMP, Assistant Secretary of War Louis Johnson said, "We have 28 different makes and models of passenger cars and 143 makes and models of trucks. In a major emergency, the maintenance of this number of different vehicles in the theater of operations obviously would be an impossible task."

In a report titled *Commercial Vehicles in the Military Service: The Problem of the Quartermaster Corps* it was argued that once payload capacity and chassis types were selected, commercial vehicles, if factory-prepared with adequate ground clearance and all-wheel drive, could be modified by the Army, post-delivery, to add brush guards, military bumpers, adjustable windshields, etc.

On 1 June 1939 the Quartermaster Corps Technical Committee recommended to the Chief of Staff that the army adopt five weight classes of trucks: 1/2-ton, 1 1/2-ton, 2 1/2-ton, 4-ton, and 7 1/2-ton. All tactical vehicles were to be all-wheel drive. Further, Quartermaster General Louis H. Bash, long an advocate of standardization, recommended that trucks be purchased in large lots, rather than small contracts, and that the maximum interchangeability of parts be a goal. Acting Chief of Staff George C. Marshall approved these recommendations shortly thereafter.

On 22 November 1939, contract W-398-QM-7471 for 3,936 1 ½-ton 4x4 trucks was issued – to Dodge. The production pilot of these trucks was shipped 16 December 1939. The pilot was approved 5 January 1940, and production began 15 January. Shipment of this order was completed on 11 March 1940. Dodge, through its Fargo Division, was awarded a second contract, W-398-QM-7813, on 16 March 1940, that contract for 2,534 1 ½-ton 4x4 trucks. Shipment of this order was completed 13 August 1940.

In mid-1940, new tentative specifications USA-LP-91-701G, had been approved and the project was moving toward the bid process. These specifications superseded tentative specifications USA-LP-91-701F, which had been approved 12 January 1940. Quartermaster documents dating to 6 July 1940 state, concerning these bids, "Bid will be opened on July 26, 1940. This will be a price test, and in the case the low bid is not satisfactory consideration will be given to negotiation. Pilot model will be required after award."

At this point, the question of who would produce the new vehicles was far from settled. On 12 July 1940, Col. H.J. Lawes called R. L. Biggers, president of the Fargo Division of Chrysler Corporation, about the specifications and bid procedure on the 1½-ton trucks. To this point, Fargo had supplied the Army with the bulk of their trucks in this size range, and in fact were producing 1½-ton 4x4 trucks for the Army when the request for bids was issued. However, Fargo was also aggressively pursuing the market for ½-ton 4x4 trucks. The smaller size had outperformed the 1½-ton 4x4 during testing in 1938.

Eight days after the bids were solicited, Chevrolet responded with a proposal, and on 6 August was declared the winner of this round, with contract W-398-QM-8278 being issued.

These trucks were considered 1941 model year trucks, and were given Chevrolet engineering symbols (sometimes referred to as model numbers) in the G-41xx range, such as G-4112, G-4103, G-4174, etc. When production of the 1942 model year trucks began, the model designation was changed to G-71xx series numbers, such as G-7103, G-7117, etc. From the standpoint of Chevrolet, the 1942 model year lasted from September 1941 until the 1945 end of production of these trucks. In the case of the engineering symbols, evidence suggests that the "G-" prefix was assigned so as to designate "government."

Chevrolet records reflect that work on the design of the truck began on 1 June 1940.

The first of these trucks was shipped by Chevrolet on October 16, 1940. In fact, during October 1940 shipments of 1½-ton 4x4 trucks were made by no fewer than eight Chevrolet assembly plants. By the time G-506 production ended in 1945, Chevrolet used the partial facilities of nine assembly plants as well as numerous other plants to produce components that were fed to each assembly plant.

The G-506 (and G-85) by the numbers

Chevrolet assigned each of their assembly plants a plant number code, which was used as the first digit in the manufacturer's serial number assigned to the vehicle (all of these trucks were produced long before today's standardized 17-digit Vehicle Identification Number format was prescribed by the Federal Government).

Those codes were:

Plant Number	Location
1	Flint, MI
2	Tarrytown, NY
3	St. Louis, MO
5	Kansas City, KS
6	Oakland, CA
8	Atlanta, GA
9	Norwood, OH
14	Baltimore, MD
21	Janesville, WI
	Bloomfield, NJ boxing plant

Chevrolet G-506/G-85 production at all of these plants except Bloomfield was begun in October 1940. By August 1943, only Atlanta and Norwood were still building these trucks.

Following the plant code in the serial number is a code which identifies the body style. The body code for first government contract vehicles built prior to September 1941 is two alphabetical characters, beginning with Y. When production of trucks on subsequent contracts was begun in April 1941 and running into September 1941, the beginning character changed to Z.

However, when additional contracts were let in August 1941, the body codes started to begin with the letter N, and that pattern remained in place through the end of production.

The two-digit body code was followed by a month code, which can be one or two digits.

For the 1941 model year production vehicles, these numbers correspond to the normal numbering of calendar months (5 is May, 12 December, etc.). That changed in 1942, when the month date code was no longer tied to the calendar year, but instead related to the vehicle model year, which began in August 1941. Thus, a vehicle with a serial number prefix of 9NJ8- (indicating August 1941 production), is older than a truck with a 9NJ1- prefix, indicating January 1942 production.

The balance of the G-506 production followed this pattern, as they were all considered 1942 model year trucks. Thus, 9NJ48- indicates that the truck was assembled in December 1944.

The serial number suffix begins with 1001, for each plant. A Chevrolet Central Office instruction letter, sent to a myriad of Chevrolet facilities concerning vehicle serial numbers states, "All G7100 Series will be numbered in sequence regardless of model starting with 1001."

Chevrolet Vehicle Assembly Plants

St Louis, Missouri

The Chevrolet plant in St. Louis was the company's largest at the outbreak of the war, encompassing 1,573,260 square feet of space on over 58 acres of land. Production of the G-506-style trucks began at this plant in early 1942. Little known is that this plant also produced GMC CCKW 6x6 trucks under contract from GMC. In fact, both the 4x4 and 6x6 were built on the same assembly line until 30 April 1942, when 4x4 production was discontinued. While not much documentation survives, it is interesting to note that surviving documents indicate that St. Louis and Baltimore were both scheduled to produce 50 G-506 trucks per day, or 1,000 per month, in late 1941.

Production of the 1941 model year trucks ended at the St. Louis plant on 30 July 1941 when serial number 3ZP07-21405 was completed. Production of the 1942-style G-506 began in St. Louis on 11 September 1941 with 145-inch wheelbase cargo truck without winch, serial number 3NJ09-1001.

Production of the G-506 continued in St. Louis until 30 April 1942, when panel truck 3NG04-5472 rolled off the assembly line. Thereafter the truck line was devoted to CCKW and DUKW production.

Baltimore, Maryland

The Chevrolet-Baltimore facility is another in which the Chevrolet 4x4 was produced on the same assembly line as was the GMC CCKW – the latter again being under subcontract to GMC. Baltimore, like St. Louis, produced their own cabs for the trucks that were assembled there. Some of the other Chevrolet G-506 assembly plants relied on Chevrolet – Flint, or Chevrolet – Commercial Body in Indianapolis for their cabs. Ultimately, both G-506 and CCKW production in the 750,380-square foot Baltimore plant was brief.

At a Chevrolet plant, probably Kansas City, G-506 Chevrolets are lined up, awaiting dispersal. The four trucks in the foreground bear registration numbers, left to right, W-325741, W-325742, W-325743, and W-325740. These particular trucks were produced at the Detroit plant under contract WQ-398-QM-8278, and the Army accepted them in 1940. General Motors LLC

Baltimore-built 4x4s were specialized trucks, such as cab-over-engine, bomb trucks and panel vans. Some of these were delivered locally to Camp Holabird. The Baltimore plant had no boxing facility for trucks, so that part of Baltimore production destined for overseas shipment were sent to the Chevrolet-Bloomfield plant for boxing. Wartime vehicle production at Baltimore was short-lived, as the plant was transferred to GM's Eastern Aircraft Division for TBM Avenger fuselage production. This production was being undertaken in the former Fisher Body plant, separated from the Chevrolet plant only by a brick wall. Material shortages forced the end of CCKW production in Baltimore, leaving only the G-506 trucks in production there. Production of the 1941 model year trucks ended at the Baltimore plant on 30 July 1941 when serial number 14ZP07-17193 was completed. Production of the 1942-style G-506 began in Baltimore on 11 September 1941 with 145-inch wheelbase cargo truck without winch, serial number 14NJ09-1001.

In November 1942 the decision was made to convert from Chevrolet trucks to Eastern Aircraft bomber component production. The conversion began on 5 February 1943 and was complete on 1 April. The final G-506 built in Baltimore was a panel van, serial number 14NG22-2176, built 16 October 1942.

Oakland, California

On the west coast, the Chevrolet plant in Oakland took part in the production of G-506 4x4s. The truck plant was located about 3 miles from the car plant, and in 1941 Chevrolet purchased additional property adjacent to the truck plant in order to set up an overseas boxing operation. However, many of the Oakland plant's products, primarily cargo and dump trucks, with a limited number of tractors, were supplied to Army installations on the west coast. Production of the 1941 model year trucks was completed at the Oakland plant on 30 July 1941 when serial number 6ZP07-10261 was assembled. The first 1942-style

Chevrolet's North Tarrytown, New York, factory was involved in the production of G-506 Chevrolets. Here, along the banks of the Hudson River, numerous 1 1/2-ton 4x4 trucks are stored in an area enclosed by security fencing. General Motors LLC

G-506 production in Oakland was on 19 September 1941 with 145-inch wheelbase cargo truck without winch, serial number 6NJ09-1001. Oakland assembled the G-506 through February 1943. The final Oakland-built G-506 was a cab and chassis, serial number 6NE14-6784, but 26 February 1943.

Norwood, Ohio

Chevrolet's 631,630 square-foot plant in Norwood, Ohio plant had the capacity to assemble 15 4x4 trucks per hour. The plant produced cargo and dump trucks, panel vans, chassis for earth boring trucks, truck-tractors and plain cabs and chassis. This production took just over half of the plant's space, the remainder being leased to Delco, who built aircraft landing gear in the plant.

Production of the 1941 model year trucks ended at the St. Louis plant on 30 July 1941 when serial number 9ZX07-12525 was completed. Production of the 1942-style G-506 began in Norwood on 16 September 1941 with 145-inch wheelbase cargo truck without winch, serial number 9NJ09-1001.

Tarrytown, New York

In Tarrytown, New York, Eastern Aircraft was building the wings and front sections of TBM Avenger aircraft in the former Fisher Body plant adjacent to the Chevrolet assembly plant. The aircraft components were supplied to Eastern's West Trenton, NJ facility, where the complete Avenger was assembled. In 1942 it was estimated that the Tarrytown subassembly operation would require an additional 240,000 square feet of space.

At first, a reduction in Tarrytown Chevrolet production was planned, which would free 90,000 square feet of

In a view of part of the production plant at North Tarrytown, Chevrolet 1 1/2-ton 4x4 vehicles are visible behind the security fence at the far left. Chevrolet North Tarrytown built these vehicles under five separate contracts during World War II. General Motors LLC

space for use by Eastern. Shortly thereafter, the decision was made to discontinue 1½-ton 4x4 production at Tarrytown entirely, the space savings satisfying the requirements of Eastern.

Production of the 1941 model year trucks ended at the Tarrytown plant on 26 July 1941 when serial number 2ZP07-20133 was completed. Production of the 1942-pattern G-506 began in Tarrytown on 11 September 1941, with 145-inch wheelbase cargo truck with winch, serial number 2NM09-1001.

On 23 December 1942, Chevrolet employees began dismantling the G-506 production equipment in Tarrytown. Although the 4x4 production was completed in December, the final shipment did not leave the facility until January 1943. The last Tarrytown-built G-506 was a winch-equipped dump truck, serial number 2NL24-7919, assembled 22 December 1942.

Janesville, Wisconsin

The Chevrolet plant in Janesville, Wisconsin was an 853,090 square foot facility when G-506 production began at the facility in October 1940. By the end of the year 333 of the trucks had been assembled at the plant, with a further 74 being completed in January 1941. Production of the 1941 model year trucks ended at the Janesville plant on 24 January 1941 when serial number 21ZP01-5445 was completed. However, it would be March 1941 before all of those trucks were shipped. There is no evidence that any further G-506 production occurred at the plant, control of which was transferred to GM's Oldsmobile Division. Beginning in August 1942, and for the remainder of the war, Oldsmobile produced 16 million 105mm, 90mm, 3-inch, and 4.7-inch artillery shells in the plant.

Completed Chevrolet 1 1/2-ton 4x4 Cargo Trucks are assembled at the Kansas City plant. The vehicles in the foreground exhibit registration numbers pertaining to contract W-398-QM-8278, with acceptance in 1941. Those numbers are, left to right, QW-326924, W-326906, W-326804, and W-326923. General Motors LLC

Kansas City, Kansas

When G-506 production began, the Chevrolet plant in Kansas City was one of the company's largest. It included a floor area of 771,240 square feet, and at the time had a capacity of 13,500 cars and trucks per month. Military 4x4 production there began with 11 trucks in October 1940, and concluded in February 1942.

Production of the 1941 model year trucks ended at the Kansas City plant on 30 July 1941 when serial number 5ZP07-13170 was completed. Production of the 1942-style G-506 began in Kansas City on 11 September 1941 with 145-inch wheelbase cargo truck without winch, serial number 5NJ09-1001. The final G-506 off the line was a 145-inch wheelbase cargo truck, without winch, serial number 5NJ02-2812.

Atlanta, Georgia

Production of the G-506 began in the 621,6300-square foot Chevrolet Atlanta plant with a single truck produced in October 1940. Production of the 1941 model year trucks ended at the Atlanta plant on 28 July 1941 when serial number 8ZP07-14532 was completed. Production of the 1½-ton 4x4 in increased steadily through February 1942, when production of the trucks ceased in Atlanta. That truck was a 145-inch wheelbase cargo truck, without winch, serial number 8NJ02-3319. G-506 production resumed at that facility in June 1943. Production of the 1942-style G-506 began in Atlanta on 15 September 1941 with 145-inch wheelbase cargo truck without winch, serial number 8NJ09-1001.

Chapter 1

The 145-Inch Wheelbase Cargo Trucks

While on 25 July 1942, responsibility for procurement of all types of Army ground vehicles was shifted to the Ordnance Department, such was not the case during the 1930s. At that time, the Army Ordnance Department had charge of the combat vehicles, while responsibility for the general purpose vehicles fell to the Quartermaster Corps. General purpose vehicles were further subdivided into two groups, administrative vehicles and tactical vehicles. Administrative vehicles tended to be commercially available vehicles, the type one would see on a dealer's showroom, or routinely on a factory assembly line. Tactical vehicles, owing to their intended field use, as a rule were equipped with all-wheel drive, a feature not commonly seen in the civilian marketplace at that time.

Because the cargo truck was by far the bulk of the production of the G-506 trucks, with the Office Chief of Ordnance – Detroit document "Recapitulation of Facility Expansions for Tracked & Wheeled Vehicles from 1940 to July 15, 1945" recording 112,938 examples having been delivered to the US government, it is this configuration that we will first examine.

Major General E. B. Gregory, Quartermaster General, spoke at the annual meeting of the Society of Automotive Engineers in Detroit in January 1941. In his remarks at that time, he stated that the Army had 60,000 motor vehicles in service, but that number would climb to 100,000 by March, and 190,000 by 1 July. By fall, he expected to have a quarter -million vehicles in inventory, with 44,000 of them being 1½-ton trucks.

The basic cargo truck is a 145-inch wheelbase all-wheel drive truck powered by a 235.5 cubic inch displacement 6-cylinder, valve in head Chevrolet engine. The initial production trucks used engines designated series AV-, while later models used BV-series engines. The 93 horsepower developed by the engine were transmitted through a four-speed transmission and two-speed transfer case. The transfer case also permitted disengagement of the front axle drive. Those axles were designed and produced by Chevrolet and had 6.67:1 differential ratios. Tires were 7.50-20 8-ply, with dual rear wheels. The trucks were equipped with a 6-volt electrical system. The cargo bed was 70 by 108 inches.

The first of the cargo trucks produced were the G-4112 trucks, with the serial prefix of YP. Winch-equipped trucks were assigned the book symbol YM, while non-winch trucks were book symbol YJ. While the Y-series trucks differed little from the April through September 1941 Z-series trucks, both types are very different from the post-September 1941 N-series vehicles.

Both the Y-series and Z-series trucks had "Chevrolet" embossed in the engine compartment side panels. The dashboard, including the instrument cluster, were taken over from the Chevrolet truck line, with few modifications. The rear cab window was held in place by screws and a diecast retainer on the outside. The window featured a mesh guard.

Contract W-398-QM-8278

On 10 July 1940, Lt. Col. J.H. Johnson of the Quartermaster Corps sought approval from the Assistant Secretary of War to initiate a 15-day advertisement for bids for the new Light Chassis (LC) trucks. In requesting approval Lt. Col. Johnson wrote, "....the following requirements for Trucks, 1½-ton (LC), 4x4, in accordance with U.S. Army Tentative Specification USA-LP-91-701G, dated July 10, 1940:

3751	Trucks, 1½-ton (LC), 4x4, Cargo
1984	Trucks, 1½-ton (LC), 4x4, Cargo w/Winch
933	Trucks, 1½-ton (LC), 4x4, Dump
284	Trucks, 1½-ton (LC), 4x4, Cargo w/Winch
90	Trucks, 1½-ton (LC), 4x4, 15' Cargo.
7042 – Total	

"2. These vehicles are to be procured for the Regular Army and National Guard under Fiscal Year 1941 Procurement Program. The Invitation for Bids will contain a fifty

Above: *The initial-production Chevrolet 1 1/2-ton 4x4 Cargo Trucks were produced under contract W-398-QM-8278, dated 6 August 1940. Production on these trucks began on 15 October 1940. This was the first wave of the Standard Nomenclature List's group G-506 and initially was Chevrolet model G-4112, revised to G-7107 in 1942. All of these were equipped with steel cargo bodies manufactured by Budd Manufacturing Co. Early on, the grille included two central, vertical elements positioned side by side, each of which was formed from a continuous rod bent back and forth. Later, this was replaced by individual vertical bars.* General Motors LLC

Below: *The following sequence of photos were taken by a General Motors photographer on 21 October 1940, six days after production began. Thus, this truck is either a pilot, or an exceptionally early production vehicle. In any event, it lacks the "Chevrolet" embossing on the engine compartment side panel that is characteristic of most early trucks. However, this truck does have six vertical ventilation louvers toward the rear of the engine compartment side panel, as well as six louvers on the panel below the front fender. These louvers remained standard throughout G-506 production. Most G-506 cargo trucks, including this one, had a 145-inch wheelbase, although long-wheelbase (175-inch) versions were built. The short-wheelbase trucks had five stake sockets on each side of the body, while long-wheelbase trucks had seven.* General Motors LLC

G-506 cargo trucks were issued with front and rear mudguards for the dual rear tires, but sometimes these flaps were removed or came up missing. (Dump trucks lacked the mudguards.) Each metal mudguard had two vertical V-shaped channels embossed on the surface for extra strength, and had two braces. General Motors LLC

percent (50%) increase option clause, to be exercised within ninety (90) days after the date of award, and the following performance requirements:

Production:	To begin not later than 75 days after date of award.
Rate of Delivery:	Not less than 700 units each 30 days thereafter.
Complete Delivery:	Within 240 days after the date of award."

On 11 July 1940, Chevrolet's Central Office received this invitation to bid. Eight days later Chevrolet returned their bid number 2-1765 to the government, and on 6 August 1940 contract W-398-QM-8278 was awarded for the production of these vehicles. The 145-inch wheelbase trucks, regardless of configuration, were assigned Chevrolet serial number prefix YP, with Chevrolet Master Book symbols YJ being assigned to the 145-inch wheelbase cargo trucks without winch, and YM for those with winch.

As noted above, Invitation for Bid which resulted in contract W-398-QM-8278 contained a clause which allowed the Government to increase the quantities of trucks on order by up to 50% beyond the initial 7,042, if invoked within 90 days of the 6 August award date. Clearly, the government exercised this option, as the *Summary Report of Acceptances Tank-Automotive Material* records indicates that 10,629 vehicles were delivered on this contract.

Because these trucks were not stock civilian trucks, and were not ordered through the dealer network, the orders flowed through the Chevrolet – Central Office, located in the General Motors Building in Detroit. Further, the trucks, not surprisingly, had components that were not listed as regular production options (RPO), so the order was entered on a Central Office Production Order (COPO).

The fuel filler is sandwiched in between the rear of the cab and the front of the cargo body in this photo of an initial-production G-4112/G-7107. The fuel tank was mounted on two supports between the chassis rails below the forward end of the body. Those supports were fastened to the right rail and are visible here. All of the tires were eight-ply, size 7.50-20. Civilian-type service headlights with bullet-shaped housings were on the front fenders. The battery, which soon would be in a battery box prominently located on the exterior of the right side of the chassis frame, was located for now under the right side of the cab floor. General Motors LLC

The engines powering these trucks were Flint-built 235.5 cubic inch 6-cylinder motors, part number 3660477, featuring "jet lubrication." The engine serial numbers for the 1941-built trucks all begin with "AV." As with the trucks themselves, the engines were serial numbered incrementally, regardless of model, beginning with 1001.

The trucks built on contract W-398-QM-8278 featured a battery mounted beneath the floorboard on the passenger's side of the cab, and was accessible through a bolted-in plate. Among the Chevrolet military 4x4s, this battery placement was unique; all the later production vehicles had a battery box mounted outboard of the passenger's side frame rail near the back of the cab.

Production of these vehicles began on 15 October 1940, with production being scattered among the various Chevrolet plants previously listed. This was followed by an invitation from the Navy to supply that branch with the 1½-ton 4x4. This invitation was received on 25 November 1940, and Chevrolet responded on 16 December, with contract NO (Naval Ordnance) 81721, for $172,000, being awarded on 10 February 1941. Research has not yet uncovered the body configuration of these trucks.

Contract W-398-QM-9402

The Army solicited an additional bid for cargo trucks on 20 December 1940. This time, Chevrolet was ready, responding with a proposal the next day. On 23 January 1941 contract W-398-QM-9402 was issued, calling for 7,004 cargo trucks with winch, 200 of which were earmarked for Defense Aid. These trucks, like their predecessors, were 1941 year-model trucks, however now they bore the serial number prefix of ZP. The Chevrolet Master Book symbol for these vehicles was ZM. Production of these trucks was undertaken at Tarrytown, St. Louis,

The G-506 cargo trucks, as seen on this initial-production example, had a recessed toolbox on the rear cross-sill, below the tailgate: a feature that was not present on the G-506 dump trucks. Two steps formed from bent rod were bolted to the top of the tailgate. A reflector is on the rear of the body to each side of the tailgate; below the tailgate are two bumperettes, a tow pintle, an electrical socket, and two tail lights. General Motors LLC

An initial-production G-506 cargo truck is viewed from the front. The small blackout marker lamps on the outboard side of each service headlight are of the early louvered style. Later vehicles used a different style black out marker lamp assembly. The front and rear axles were the "banjo" type. The driver's rear-view mirror was mounted on a telescoping arm, and the tow hooks cant inward, a characteristic of the early trucks. General Motors LLC

The interior of this very early Chevy reveals some of the features of trucks on the first contracts. The dash has civilian gauges, a key ignition and appears to have been painted with"wrinkle finish" OD paint. In the center of the dash under the control knobs is an ash tray found only in the earliest of these vehicles. The floor has a Jute-backed rubber floor mat with a rubber "bellows" type seal under the starter button. The seat is covered in an olive green leatherette covering and door inner seals are also OD green. The knob directly under the dash is for opening the fresh air ventilator which is located on the top of the cowl. Also interesting to note is the gloss black steering column and data plates on the glove box door with a chrome latch button. General Motors LLC

Another initial-production Chevrolet 1 1/2-ton 4x4 Cargo Truck produced under contract W-398-QM-8278 and accepted in 1940 or 1941 is viewed from the left side. The spare tire was stored horizontally under the body. Below the spare tire is the exhaust. The closed cab would remain a feature of the G-506 cargo and dump trucks throughout production. TACOM LCMC History Office

Oakland, Janesville, Norwood, Atlanta, Kansas City and Baltimore. While production halted in Janesville in January 1941, it continued at the other facilities until near the end of July. These trucks were nearly identical to the first-contract trucks (the YP). However, there were some differences. Notably, the parking brake on these (and subsequent) cargo trucks was a band around a drum affixed to the transfer case. The YP trucks had cables running to the rear brake drums, manually applying the rear service brakes for parking. In fact, Chevrolet Instruction Letter #41-47 of 10 March 1941 implies that this is the defining difference between the Y and Z prefix trucks. A further change to the brake system was the introduction of a power brake system, utilizing a Bendix Hydrovac unit. The system introduced on the Z prefix trucks used the dual piston Bendix model H-66-33, part number 372290 Hydrovac, Chevrolet part number 606446.

Additionally, the battery was relocated from beneath an access panel in the floorboard to a frame-mounted battery box. The right-side running board was shortened and the right rear bottom corner of the cab was redesigned so as to provide access to the new battery location.

Evidence suggests that the bulk of the ZP series trucks utilized engine part number 839709, the serial numbers of which began "AVB." Again, the engines were serial numbered incrementally, regardless of model, beginning with 1001.

The Quartermaster Corps continued to argue for standardization of vehicles, feeling that any disparity in pricing through negotiation rather than competitive bidding would be outweighed by the simplified logistics inherent with a commonality of parts.

The Government publication *US Army in WWII - The Army and Economic Mobilization* records: "The Quartermaster Corps continued to use advertising in

This 1941 Chevrolet model G-4163 1 1/2-ton 4x4 Cargo Truck with Winch, registration number W-332065, was produced under contract W-398-QM-9402 dated 23 January 1941. On this vehicle the battery had been moved from below the right side of the cab floor to a battery box attached to the right side of the chassis frame. The lower corner of the cab and the rear of the apron (the vertical panel above the running board) were modified to provide clearance for the battery. General Motors LLC

the placement of vehicle contracts until February 1941, when OASW (Office Assistant Secretary of War) ordered its discontinuance:

'On February 4, 1941 this office received instruction from the Under Secretary of War to suspend, until further notice, all public advertising for motor vehicle equipment for the defense program. Such procurement as has been accomplished since this suspension has been handled by securing informal quotations from manufacturers who have previously furnished satisfactory equipment of the various types needed.'"

Contracts W-398-QM-10328 and DA-W-398-QM-8

Two contracts were initially issued for production of the 1942 year model trucks, W-398-QM-10328, which would be renumbered W-374-ORD-2726 following the July 1942 Ordnance Department take over of vehicle procurement, and Defense Aid contract DA-W-398-QM-8. The invitation for bids for both of these contracts had been issued on 6 May 1941, with Chevrolet responding on 14 May. The contracts were issued on 7 June, and included cargo trucks with and without winch, long wheelbase cargo trucks, dump trucks with and without winch, tractors, airfield crash trucks and cab and chassis. The plant code of serial numbers of these trucks was followed by "N" rather than the Z used on the previously produced trucks. Cargo trucks had the codes NJ for the trucks without winch, and NM for the trucks with winch.

The eight varieties of trucks on contract W-398-QM-10328 were handled by Chevrolet Central Office Production Orders 42-1X through 42-7X. Defense Aid contract DA-W-398-QM-8 was covered by Chevrolet Central Office Production Orders 42-63X through 42-68X.

Production of the 1942 model-year 1½-ton Chevrolet 4x4 began 11 September 1941 at Tarrytown, St. Louis,

Chevrolet model G-4163 W-332065 is viewed from the front. The bottom center of the front bumper was cut out to accommodate a winch cable roller. In the top center of the bumper is a guide for the starting crank. A round reflector is visible on each side of the front of the cargo body. Also in view are the two top-mounted hinges of the windshield and the windshield wipers. These wipers used a special articulated jointed arm to allow the blades to cover the entire windshield glass. These arms were nicknamed " Grasshoppers" due to their likeness to that insect's rear legs. General Motors LLC

The same truck, W-332065, is seen from the rear. The taillights were recessed in the two round holes below the tailgate. A padlock is attached to the toolbox door on the rear cross-sill below the tailgate. Tailgate chains have a OD colored leatherette covering and a rear cargo compartment safety strap is attached the side racks. General Motors LLC

Kansas City and Baltimore. These were followed by Atlanta on 15 September, Norwood the next day, and Oakland on 19 September. Production of the 1942 model year trucks would continue to the end of the war – which would, as will be seen later, create havoc of Chevrolet's date code system.

It was at this time these trucks received the most sweeping changes incurred during their production. The electrical system switched from a 6-volt positive ground system to a negative ground system of the same voltage. The fuel filler was changed – first to 2 1/16 across ears (3-inches across cap – a 1/8-inch increase over the YP and ZP cap), and then later following a 9 December 1941 directive from Lt. Col. Georges F. Doriot to modify the gas filler to accommodate a new gas can, the cap was enlarged to 4 21/32-inches outside diameter.

Another subtle external change was the tow hooks atop the front bumper. Earlier trucks had canted tow hooks held in place by two bolts. The 1942-style tow hooks were not canted and were mounted with three bolts. Inside the cab, the civilian-style instrument cluster gave way to a very utilitarian military style gauge cluster, a nod to the government's feeble wartime attempt at standardization.

Under the hood, all of the conventional cab G-7100 series series trucks used engine part number 839821. These engines were built in Flint, and had serial numbers beginning "BV." Like the trucks themselves, all the BV engines built for the remainder of the war were considered by Chevrolet to be "1942" engines.

Chevrolet engine serial numbers for engines considered 1942 model began with 1001. The Flint engine plant assigned serial numbers in sequence, regardless of the engine model. That is, two sequential engine numbers were not necessarily the same model or displacement engine.

The exception to these rules were the 270 cubic inch engines built under contract for GMC for use in CCKWs. These engines had their own block of nine-digit serial numbers beginning with "270."

In the event that a 1941 engine was factory installed in a 1942 model truck, a prefix of "2" was to be stamped preceding the regular 1941 prefix.

After 2,045 cargo trucks had been produced on contract W-298-QM-10328, and 783 cargo trucks with winch for contract DA-W-398-QM-8, the latter being built in Tarrytown, were completed, the method of rear window installation changed. The screwed-on diecast molding was eliminated and a retainer used instead.

Up to this point, all of the 145-inch wheelbase cargo trucks had been equipped with all steel beds made by the Edward G. Budd Manufacturing Company. The Budd Company had been contracted by Chevrolet to furnish 70" x 108" steel cargo bodies for the following contracts, in the noted quantities:

Qty	Contract Number	Approximate Date of Order
8517	W-398-QM-8278	October 1940
6769	W-398-QM-9402	March 1941
8291	W-398-QM-10328	2 July 1941
2524	DAW-398-QM-8	November 1941
2851	W-398-QM-10939	November 1941
19488	W-398-QM-11510	November 1941

However, once the United States was fully involved in the war, the demand for raw materials, including solder and steel, reached critical levels. Shortages ensued, which impacted all manufacturers.

On 1 May 1942 the Studebaker Corporation shipped to Lt. Col. William Johnson at the Holabird Quartermaster Depot one of their closed cabs that lacked the detailed finishing typically found on vehicle cabs, notably the soldered body seams. In the letter accompanying the cab, Studebaker Chief Engineer Jeffries stated that omitting these steps would save two pounds of war-critical solder per cab, as well as a considerable amount of labor, yet another item in short supply. Accordingly, approval was sought from the Quartermaster to discontinue these finishing operations. Johnson no doubt approved this change, as most truck manufacturers dropped these operations the next month.

To lessen the demand, the War Department determined that the cargo bodies of most tactical trucks would be produced of wood rather than steel. Work had begun on developing wooden bodies for various military trucks as early as 28 March 1942. As demand for steel increased, this change became more critical until the Iron and Steel Limitation Order as amended 13 July 1942 mandated that 60 days after the governing date of 15 July the use of iron or steel would be prohibited for truck and trailer bodies on orders for Army, Navy and Maritime Commission, except tank and dump bodies, as well as essential hardware, structural and bracing members for bodies essentially of wooden construction. In May 1942, the Holabird Quartermaster Base issued

This factory photo and the one that follows were taken to document several places where the paint cracked and spalled off of the front clip of a Chevrolet 1 1/2-ton 4x4 Truck. Those areas are circled in white. This vehicle has the later standard type blackout marker lights and the steering wheel is now an Olive Green plastic one. This vehicle is probably for export as it has only a single windshield wiper on the drivers side General Motors LLC

Tentative Specification ES-No.754 for the 70" x 108" wooden cargo bed.

In a telegram on 25 May 1942 it was announced that a deadline of 1 August had been set to achieve 50% wooden cargo body use, with 100% mandated by 1 September. This directive, although overly optimistic, gives an indication as to the sense of urgency to make this change.

In complying with the mandate of wooden cargo bodies, 5,425 wooden 70" x 108" cargo bodies were ordered by Chevrolet from Budd for use in 1942, with deliveries beginning in August.

Production of the 1½-ton 4x4 Chevrolet with wooden bed began in August 1942, with both wooden and steel bodies being used in August, September and October of that year. Following company-wide inventory of assembly plants during the last week of October 1942, all wood cargo beds began to be used exclusively. During August, September and October, the Chevrolet serial numbers of trucks equipped with wooden cargo bodies had a W suffix added.

As noted above, initially Chevrolet and other truck manufacturers were initially responsible for building or buying the wooden cargo bodies required to fill their orders. However, that would change in 1943. A December 1942 memo from Colonel Tossy of the Detroit Ordnance District advised that this practice was going to be discontinued. The memo explained:

"The reason for the above is that, in the future the Government will do all contracting for your wood cargo bodies."

"It is apparent that the future program will give consideration to subcontractors, such as those who are presently supplying you with dimensional lumber, parts, and other material used in the assembly of the bodies, in that they will be utilized, as far as possible, as suppliers to the body manufacturers whom the government will select. Also, it is the intent of the Government that present commitments to the prime vehicle contractors will be recommitted to the body builders to the maximum extent, and that all subcontractors will be protected as far as possible. The exception to the above will be those subcontractors who will eventually be eliminated from the wood body program because they have other business."

In accordance with this directive, beginning on 1 June 1943 cargo trucks without winch were built with government-furnished bodies, and the same held true for cargo trucks without winch built after 1 September 1943. Chevrolet noted that the serial numbers impacted by this directive were NJ19 and high and NM21 and higher.

A total of 4,417 non-winch cargo trucks were delivered against this contract in 1941, with a further 2,372 examples in 1942, and the final 838 of the non-winch trucks in early 1943. Delivery of winch-equipped trucks on this order was accomplished in only two years, with 441 in 1941 and 218 in 1942.

Contract W-398-QM-11510

While the 7,086 cargo trucks without winch and 1,200 trucks with winch initially ordered on the June 1941 contract W-398-QM-10328 was certainly not insignificant, it pales in comparison to the numbers on the next major order.

The Letter Order for what would be contract W-398-QM-11510 was issued on the last day of 1941, with the contract being executed on 4 January 1942. Delivery of trucks on this order began in 1942 and would continue into January 1944. Ultimately 30,705 cargo trucks without winch and 8,650 examples with winch would be delivered under this contract.

Two spots on the right side of the hood where the paint cracked off are indicated. These two photos also provide clear details of the fenders, brush guard, headlights, and front clip. General Motors LLC

Open Cabs

For the bulk of this time frame, most US military trucks began to be built with open-top, military style cabs. The Chevrolet cargo truck, however, was an exception.

GMC's Production Control Manager Russell Crist reported that in an October 1942 meeting, "Mr. Shaw of Chevrolet reported to the Ordnance Department in regard to the status of open cabs on Chevrolet trucks as follows:

(a) 4780, 1-1/2 ton 4x4 trucks on their present contract could be equipped with open type cabs, and the obsolescence cost would be $126,000.00. In view of the high obsolescence cost Ordnance approved the use of closed cabs on these vehicles."

(b) An additional quantity of 7638 vehicles, apparently an increased quantity on their contract, could be equipped with open type cabs with no obsolescence cost involved."

A significant portion of the 1½-ton 4x4 Chevrolet production was destined for Russia under defense aid. In fact, the Soviets received 60,475 of the trucks, and owing to the harshness of the Russian winters, objected to the open cab on both the Chevrolet and the Studebaker US6 trucks.

The War Department, hoping to simultaneously satisfy the Soviet requirement for closed cabs, and the US requirement for open cabs yet minimize waste, authorized that closed cabs, originally intended for GMC

The G-506 Chevrolets were equipped with radio-suppression equipment, which controlled the radiation of radio interference generated by the vehicle's electrical equipment, which would cause static in the radio equipment of nearby radio-equipped vehicles. Radio suppression was achieved in early-production G-506s with resistor suppressors, filters, condensers, and bonding. Later-production vehicles used shielding, suppressors, and condensers to control radio interference. This photograph indicates the early-type radio-suppression features on the front end of a G-506, which included bonding straps (indicated by the number 1; 13 flexible metal bonding straps were used in total) and numerous plated bolts, nuts, and toothed lock washers. General Motors LLC

Below: *Plated bolts, nuts, and lock washers employed as radio-suppression bonding measures on the fender and the hood of a G-506 are depicted. Another bonding measure barely visible not in this photo was the affixing of metal screens inside the ventilation louvers on the side panel of the hood and the fender skirt. This truck has these screens attached by rivets, the heads of which are visible surrounding the louvers.* General Motors LLC

CCKWs, be used instead on 1½-ton Chevrolet trucks. As Chevrolet actually manufactured the CCKW closed cabs for GMC, this exchange was greatly simplified.

In response to a query in September 1943 with regard to why the Chevrolet did not feature an open cab, Captain I. E. Parsons of the Ordnance Department responded;

"1. The reason no open cab trucks are being built in the 1½-ton, 4x4 class is because practically one hundred per cent (100%) of the present production is going to the Russians and they insist on closed cabs for driver protection against the extreme cold experienced in their operations.

"2. The few trucks of this type that we use are chiefly of the truck-tractor class where one of the main advantages of the open cab, namely, low silhouette, is of no importance due to the high silhouette of the semi-trailers.

"3. Since it is not economical to produce both open and closed cabs on one vehicle type and since our requirements are so small compared to the overall production, the truck, 1½-ton, 4x4 is being produced with the closed cab exclusively.

"4. Had this vehicle been used one hundred percent (100%) by our Army, the military characteristics would have been adhered to."

Composite Beds

The demands on the lumber industry – both hard and soft wood – were so great that in 1943 there developed a wood shortage, while at the same time the steel crisis that arose in 1942 was easing.

Above: *This Chevrolet 1 1/2-ton 4x4 Cargo Truck, manufacturer's serial number 8NJ29-20271, was photographed for the Ordnance Operation, Engineering Standards Vehicle Laboratory, Detroit, on 13 July 1944. The "8" in the Ordnance number referred to Chevrolet's Atlanta plant. The "NJ" stood for 1942-1943-production cargo trucks. The number 29 indicated that the vehicle was built in May 1944. The number 20271 was the serial number, which on G-506 trucks began with 1001 in August of each year. The fuel filler neck is noticeably taller than on the initial-production G-506 trucks. This vehicle had the composite, or "bolted-up" body, the first contract for which was placed on 4 March 1944 and completed in June of that year. Four bolt heads are visible on the vertical flange at the front corner of the body, and these were a key identifying feature of the composite body.* TACOM LCMC History Office

Below: *Chevrolet 1 1/2-ton 4x4 Cargo Truck registration number 3446030 is representative of the 1944 Chevrolet model G-7107, produced under contract DA-W-398-QM-517, issued on 31 August 1942. The truck had the bolted-up composite body, as indicated by the four faintly visible bolts on the front corner of the body. The number E-2-62 was painted on the door. The letter "S" is very faintly visible on the rear of the hood: this signified that radio-interference suppression equipment on the vehicle had been tested and had passed. A new model of blackout marker lamps with small, rectangular apertures on the fronts were present, and a blackout headlight was mounted on the left fender outboard of the service headlight. An extension of the brush guard was installed to protect the blackout headlight.* TACOM LCMC History Office

The following series of photographs was part of a study and report on the effects of the impact of various weights of steel fragments on the Chevrolet Model G-7107 1 1/2-ton 4x4 Cargo Truck. The test, Department of the Army Project No. 503-05-004 and Ordnance Corps Project No. TB3-0226E, was conducted between July 1955 and September 1956 as part of a larger program to assess the vulnerability of various types of U.S. and foreign unarmored wheeled vehicles to artillery and rocket fragments. As part of this study, photographic double exposures were generated, showing the interior of the truck with the exterior of the truck superimposed, from many different angles and perspectives. In this photo dated 2 November 1956, the seeming vertical streaks are from the folds in a canvas backdrop. National Archives

In early November 1943 General Campbell sent a memo to the Chief of Ordnance raising the question, "Why are not wooden bodies replaced with steel now that steel is available?"

This inquiry set off a flurry of correspondence, which included several notations that in late November there was considerable pressure being exerted on Ordnance to introduce aluminum cargo beds, noting "...are now trying to get us to use aluminum, of which there is apparently a great surplus."

By 16 December the aluminum idea had been vetoed, citing both the considerably higher cost of aluminum as well as the difficulty in fabricating beds from the material. The text of the decision read, "...only in cases where its use is indicated for technical or tactical reasons. Specific application of aluminum to truck cargo bodies has been thoroughly considered, but such use has been decided against, primarily because the much higher cost involved could not be defended and also due to disruption in engineering work involved in such a change. No further action is being taken on this particular application nor will a request from prime contractors for such use be entertained."

However, the matter of again utilizing steel cargo beds had been resolved on 3 December 1943, when the Assistant Chief, Tank-Automotive Center, John K.

This side view of the front end of the truck illustrates the same parameters of angle of attack as the preceding photo. The steering wheel, control levers, and engine may be seen within the truck. The registration number on the hood is visible: 349931. The engine used in the G-506 trucks was the Chevrolet GV: these were assigned serial numbers starting at 1001. National Archives

Christmas, sent a memo to the Chief of Ordnance – Washington outlining the construction of future cargo beds. This memo introduced what is commonly referred to as the "composite" cargo body, a body with a wooden floor and steel sides. Beyond introducing the composite body, the memo also offered insights into wooden cargo body production, and is quoted here.

"1. At the outset of the present war effort our cargo bodes for trucks were of all-steel construction. We were directed by A.S.F. on 31 May 42, QM-451.21 to change over to wood cargo bodies at the earliest possible date and no later than 1 Sept 42 because of the shortage of carbon steel sheets. It was originally planned to have the vehicle manufacturers supply the wood cargo bodies as part of their vehicle contract. However, it was ultimately decided to place prime contracts for wood cargo bodies with small manufacturers according to a plan proffered by the Smaller War Plants Corporation."

The memo continued;

"3. Experience with the present wood cargo bodies, in which there are a minimum of steel reinforcing members, indicates the advisability of returning to the all-steel construction. However, this would adversely affect the smaller companies now producing the wood cargo bodies and consequently we propose effecting an engineering change at the earliest possible date to utilize a larger percentage of steel and secure a more rugged and durable body. The Tank-Automotive Center, working with industry, has prepared working sketches....and tentative bills of material for a composite cargo body. This body uses steel for the front panel, side panels, tail gate, bolsters and floor skid strips. Wood is used for the floor, longitudinal side sills and tool box floor."

"4. The change will bring about an increased requirement of approximately 2000 tons per month of carbon steel sheets."

"6. The Tank-Automotive Center is preparing an O.C.M. item to be read for the record. They

In a frontal view of the Chevrolet 1 1/2-ton 4x4 Cargo Truck during the fragment tests, the radiator fan, fan belt and pulleys, oil filter, and generator are visible behind the grille. Two jacks were inserted under the chassis frame. National Archives

are also arranging for the early completion of working drawings and detailed bills of material for the following body sizes: 80" x 144", 70" x 108", and 88" x 132". It is believed that this engineering change can be placed in production by approximately 1 July 1944 if we can receive prompt authorization."

In order to implement this change, beginning in March 1944 Budd supplied 70" x 108" composite cargo body material to a number of firms.

3-4-44	2228ea	Detroit Ord Dist	W-20-018-ORD-358
4-10-44	1588ea	Providence Body	W-19-020-ORD-2063
4-17-44	1230ea	Charak Furniture Co.	W-19-020-ORD-1059
4-18-44	1230ea	Old Colony Furniture	W-19-020-ORD-1170
7-3-44	4638ea	Krieger Steel Sections	NXS33610
9-12-44	900 ea	Mott Brothers	W-30-069-ORD-2183
9-14-44	2538ea	Ferguson Brothers	W-30-069-ORD-2190

In this left-hand, x-ray-like double exposure of the front end of the truck, features visible include an angle-iron brace running from the top of the radiator back to the chassis frame below the steering column. Below the brace, the generator is visible. National Archives

This ghostly double-exposure was taken from the left rear of the cab of Chevrolet 1 1/2-ton 4x4 Cargo Truck registration number 349931 in a shop at Aberdeen Proving Ground on 2 November 1956. This was a 1942-model-year truck manufactured at Chevrolet's Detroit plant under contract W-398-QM-10328. National Archives

While not all of these bodies were destined for 1½-ton 4x4 Chevrolet trucks, many of them were.

During the course of production of these trucks, the dual piston Hydrovac ceased to be installed. Instead, a single piston unit, Bendix model HS-25-25P-53, part number 372774 was used. On 14 April 1944 approval was given to add a baffle to the Hydrovac in order to prevent water entry through the Hydrovac air cleaner.

Deliveries of the 145-inch wheelbase cargo trucks ceased in June 1945. In that month, 57 of the trucks with winch were accepted, as were 904 trucks without winch. These trucks were built in Norwood, Ohio and Atlanta, GA; the Chevrolet plant in Oakland having ceased G-506 production in February 1943, and with the various other plants having halted 1½-ton 4x4 production in 1942 or prior.

Some sources indicate that G-506 cargo truck production continued for some time later than this. This belief stems from two mistakes. First is the mistaken belief that civilian truck production did not resume until after VJ-Day, which lead some to misinterpreting a photo showing both G-506 and civilian trucks sharing an assembly line.

In a view of the test truck from the rear, the fuel tank is in the foreground, between the chassis frame rails. The elbow for the fuel filler is atop the tank. National Archives

The second mistake was made by Chevrolet during the first week of January 1945. As previously mentioned, the bulk of the G-506 trucks were considered 1942-models by Chevrolet, regardless of year production. This decision meant that, at least according to the Chevrolet serial number system, 1942 had more than 12 months. Just as this has confused enthusiasts since, it also led to confusion in Atlanta – and during the first week of production in 1945, the truck serial numbers were erroneously stamped with a prefix of 8NJ49, rather than the proper 8NJ37. This mistake, which was discovered on 9 January, after the trucks had been shipped, impacted 296 vehicles, serial numbers 8NJ49-13659 through 8NJ49-13842 (registration numbers 3478400 through 3478583) and 8NJ49-13843 through 8NJ49-13954 (registration numbers 3292617 through 3292728).

Above: *A right-rear view of the cab provides a clear view of the battery holder, which has a left-pointing flange bent into each side to allow for fastening the holder to the chassis frame with two hex screws per side.* National Archives

Below: *The cab and part of the chassis frame and fuel tank are viewed from the upper right. The steering wheel and shifting levers are visible inside the cab. The cab seat was not installed in this vehicle.* National Archives

ABOVE: *Elements of the drive train are visible in this photo from above the rear of the vehicle: the Chevrolet BV engine and the transmission, the transmission-to-transfer-case drive shaft, and, mostly hidden under the frame cross member to the left of the battery, the transfer case. To the right of the transmission-to-transfer-case drive shaft are the drive shaft from the transfer case to the front axle (mostly hidden in the shadows) and the hydrovac, part of the hydraulic brake system.* National Archives

BELOW: *Chevrolet 1 1/2-ton 4x4 Cargo Truck registration number 349931 is seen from the upper left in an X-ray-like double exposure taken at Aberdeen Proving Ground on 2 November 1956 during the truck-vulnerability study.* National Archives

Above: *Much of the engine compartment is visible in this double exposure. Above the front of the exhaust manifold is the horn. To the front of the carburetor air cleaner, on top of the rocker arm cover (or valve cover) is the rocker arm cover air cleaner. To the rear of the carburetor air cleaner are the oil cleaner (left) and the regulator (right).* National Archives

Below: *The engine compartment is seen from a different perspective, with elements of the steering gear and front suspension visible adjacent to the left front tire. The radiator braces also are in view.* National Archives

Above: *In this low-angle double exposure from the front left of Chevrolet 1 1/2-ton 4x4 Cargo Truck registration number 349931, above the tire is the left front shock absorber and shock-absorber link. Jacks are supporting the front end of the truck. Folds in the canvas backdrop are the cause of the vertical streaks in the photo.* National Archives

Below: *Elements of the engine compartment, the front wheels, and the front bumper and its fittings are in view in this 2 November 1956 double exposure. A ruler has been placed on the right side of the bumper, for scale.* National Archives

Above: *An elevated frontal view of Chevrolet 1 1/2-ton 4x4 Cargo Truck registration number 349931 gives an idea of the chassis frame with reference to the cab and the front clip.* National Archives

Below: *The truck from the fuel tank forward is observed from the upper right. To the left of the transmission-to-transfer-case drive shaft is a truncated pipe: this is the exhaust, with the muffler and tailpipe removed.* National Archives

This final double-exposure from the 1956 vehicle-vulnerability study was taken from directly above the cab. The chassis frame, BV engine and accessories, front bumper, running boards, spare tire, and fuel tank are light colored, while the cab roof, hood, and front fenders are dark. National Archives

Above: *Chevrolet 1 1/2-ton 4x4 Cargo Truck manufacturer's serial number 8NJ35-6339 was photographed for the Ordnance Operation, Engineering Standards Vehicle Laboratory, Detroit, on 8 January 1944. The truck was built at the Atlanta factory in November 1943. This truck has the wooden cargo body, partially obscured by the tarpaulin. Wooden cargo bodies were introduced in late 1942 in response to a shortage of steel. A reflector was affixed to each side of the front of the forward cross-body sill. A standard vehicle type rifle rack has been installed inside the cab on the passenger side.* TACOM LCMC History Office

Below: *The same Chevrolet 1 1/2-ton 4x4 Cargo Truck shown in the preceding photo, manufacturer's serial number 8NJ35-6339, is viewed from the front right on 8 January 1944. On the top center of the front bumper was a square tab with a hole through it, for supporting the manual engine starter crank during starting operations. The battery is faintly visible on the inboard side of the five-gallon liquid container on the running board. Only the front mudflaps for the rear tires were installed on this vehicle.* TACOM LCMC History Office

Above: *In a 3/4-left-rear view of Chevrolet 1 1/2-ton 4x4 Cargo Truck 8NJ35-6339, markings are stenciled on the lower-center part of the side of the tarpaulin for "U.S.Q.M.C." (U.S. Quartermaster Corps) and the maker of the tarpaulin, the Hettick Manufacturing Co., Toledo, Ohio.* TACOM LCMC History Office

Below: *The tarpaulin has been removed from Chevrolet 1 1/2-ton 4x4 Cargo Truck 8NJ35-6339, revealing the construction of the wooden body. Stake pockets were fastened to the wooden side boards with hex screws. The steel rear sill was simplified and reduced in size compared with the sills on the previous steel cargo bodies. The pockets were braced at the bottom, and there also were two separate, diagonal braces made of channel material. Reflectors and lashing hooks were fastened to the side boards.* TACOM LCMC History Office

Above: *The same cargo truck, 8NJ35-6339, is viewed from the left with the tarpaulin removed. From this angle, the registration number is faintly visible on the hood: 3289948 S. This was a vehicle from contract W-374-ORD-2694, production order number T-6057, from the original contract dated 4 January 1942.* TACOM LCMC History Office

Below: *Chevrolet 1 1/2-ton 4x4 Cargo Truck serial number 8NJ35-6339 is viewed from the right side with the tarpaulin removed in a 5 January 1944 photograph. The fuel tank may be faintly seen below the spare tire, between the rails of the chassis frame. Note cutouts in the bottom rear corner of the cab and in the apron above the rear of the running board have been enlarged to provide access to the frame mounted fuel filter.* TACOM LCMC History Office

Above: *Five-gallon liquid containers are in the holders on the running boards in this overhead view of Chevrolet 1 1/2-ton 4x4 Cargo Truck serial number 8NJ35-6339 in a photo dated 28 January 1944. To the front of the windshield, the outline of the cowl vent is visible. On the floor of the wooden cargo body, bolted between each wooden plank are metal rub strips, whose technical name is "wear, iron, platform."* TACOM LCMC History Office

Below: *Chevrolet 1 1/2-ton 4x4 Cargo Truck registration number 3289948 and manufacturer's serial number 8NJ35-6339 is viewed from the rear in a photo dated 5 January 1944. Wooden cargo bodies retained the toolbox below the tailgate. Also below the tailgate, mounted on metal plates, are, left, the left taillight and the electrical receptacle, and, right, the right taillight. On each side of the tailgate is a reflector and a tarpaulin-lashing hook. These are the later stamped metal hooks which replaced the early cast metal ones.* TACOM LCMC History Office

Above: *The interior of the same cargo truck is viewed with the tailgate lowered. There is a cutout in the slats of the front rack to accommodate the rear window of the cab. A mesh guard is provided for that window.* TACOM LCMC History Office

Below: *A final view of Chevrolet 1 1/2-ton 4x4 Cargo Truck registration number 3289948 and manufacturer's serial number 8NJ35-6339, dated 5 January 1944, shows the vehicle from the front. The early-style "snake" grille, fabricated from steel rods bent in accordion fashion, had been replaced by grilles made from individual vertical rods attached to a frame. The type of grille shown here is referred to as asymmetrical, as it has an extension on the left side to protect the blackout headlight.* TACOM LCMC History Office

Above: *Chevrolet 1 1/2-ton 4x4 Cargo Truck with Winch registration number 394705 is the subject of this four-photo series taken at Aberdeen Proving Ground, Maryland, on 22 May 1944. This wooden-bodied truck was manufactured under contracts W-398-QM-11510 and W-374-ORD-2694. "CHEVROLET" is embossed on the side panel below the hood to the front of the louvers.* National Archives

Below: *The same cargo truck is viewed from the right rear quarter. Goodrich Silvertown non-directional, super-traction tires were mounted, size 7.50-20. Details of the cutouts in the cab and the apron to provide clearance for the battery are visible.* National Archives

Above: *Chevrolet 1 1/2-ton 4x4 Cargo Truck with Winch 394705 is viewed from the right side at Aberdeen Proving Ground in a photograph dated 22 May 1944. A six-foot measuring stick graduated in inches is positioned to the front of the vehicle, for reference.* National Archives

Below: *In this overhead view of the same cargo truck, it can be seen that some of the floor planks in the cargo body were not full-length, but were pieced together. The planks were generally of clear lumber, with at least one large knot visible.* National Archives

Above: *This 1 1/2-ton 4x4 Cargo Truck, registration number 3344921, is one of three such experimental vehicles built by Chevrolet using amixture of Chevrolet G-506 and GMC CCKW components. The experimental trucks used the GMC engine, transmission and transfer case. The side panels of the hood were the type used on G-508 CCKWs, with 12 vertical louvers, and the registration number doesn't tally with known registration numbers for Chevrolet G-506 vehicles. The vehicles were tested at Aberdeen Proving Ground, where one of the trucks was converted to truck-tractor configuration. In the left background are two Chevrolet-built T17E1 Armored Cars.* TACOM LCMC History Office

Below: *Chevrolet 1 1/2-ton Cargo Truck with Winch registration number 394705 was a 1942-model-year vehicle produced under contract W-398-QM-11510, dated 4 January 1942, and subsequently continued under contract W-374-ORD-2694. The wooden cargo body has additional vertical braces adjacent to the first, third, and fifth stake pockets.* TACOM LCMC History Office

Above: *The same cargo truck with winch shown in the preceding photograph, registration number 394705, is viewed from the right rear. A safety belt is clipped to the rear stakes. Considerable white overspray is visible around the identification stars on the tailgate and the cab door. Neither of the rear mudguards were installed on this vehicle.* TACOM LCMC History Office

Below: *On 7 November 1942 the following series of views was taken of a G-506 Chevrolet Cargo Truck with a winterization kit installed. This kit was intended to boost the performance and survivability of these trucks in arctic conditions, where extreme cold could wreak havoc with batteries, fuel systems, and crews. In this photograph, a tarpaulin designed to insulate the engine compartment is installed.* General Motors LLC

Above: *The Chevrolet 1 1/2-ton 4x4 Cargo Truck is viewed from the right side with the tarpaulin rigged over and around the forward part of the vehicle. As several heaters for vehicular components and an under-chassis heater were part of the winterization kit, the tarpaulin had the effect of keeping the heat from those devices from dispersing.* General Motors LLC

Below: *The tarpaulin did not cover the top of the cab, but it did wrap around and under the chassis to the rear of the cab. A set of ropes was provided to lash the tarpaulin in place; one rope passed around the rear tires. On each side of the tarpaulin below the cab doors was an access flap.* General Motors LLC

Above: *The winterization kit included a heating coil at the bottom of an insulated battery box. The box is shown here with a Delco-Remy Model 19-C-1 150 amp hp battery installed.* General Motors LLC

Below: *The winterization battery box is shown with the battery and its tray removed, permitting a clear view of the heating coil. Note how the battery box rested on a holder with a slanted side.* General Motors LLC

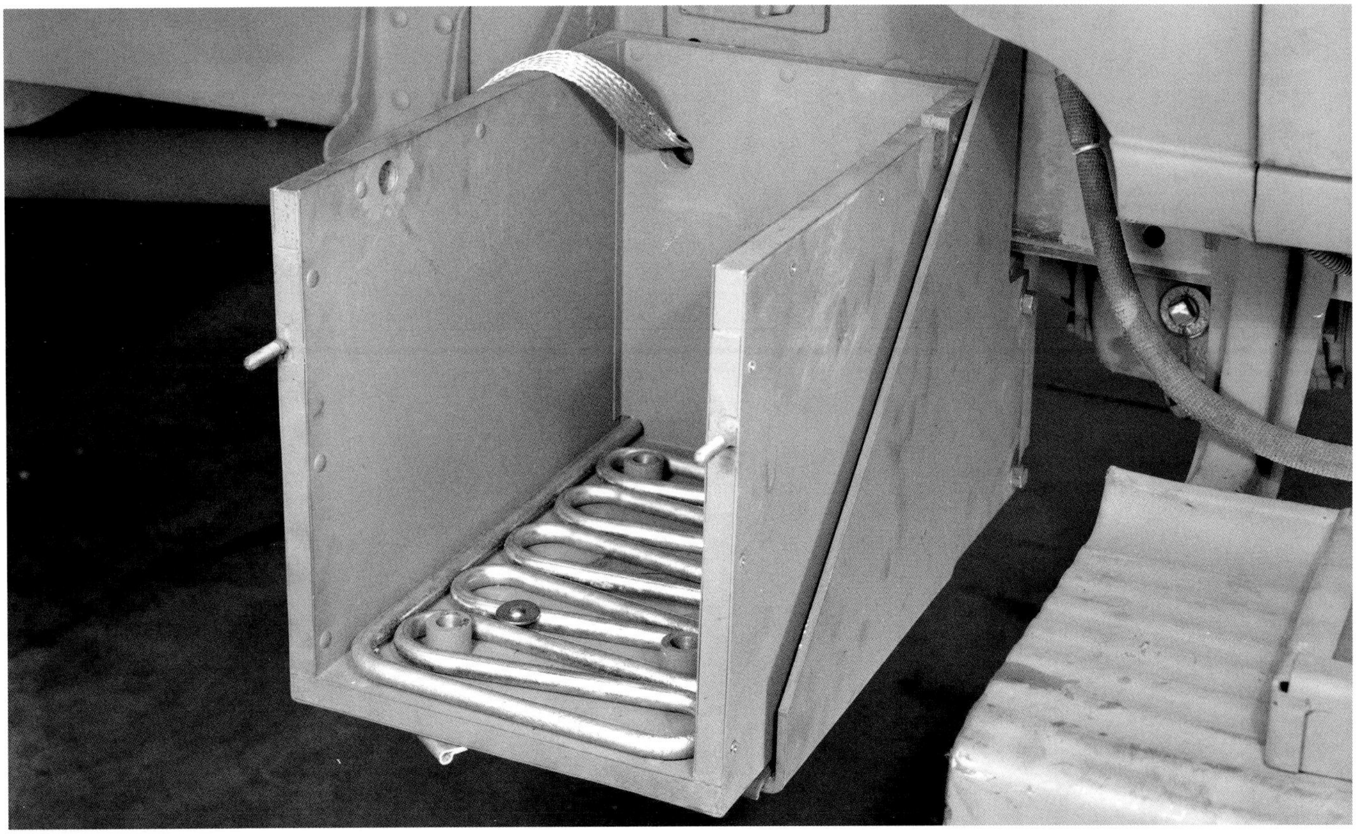

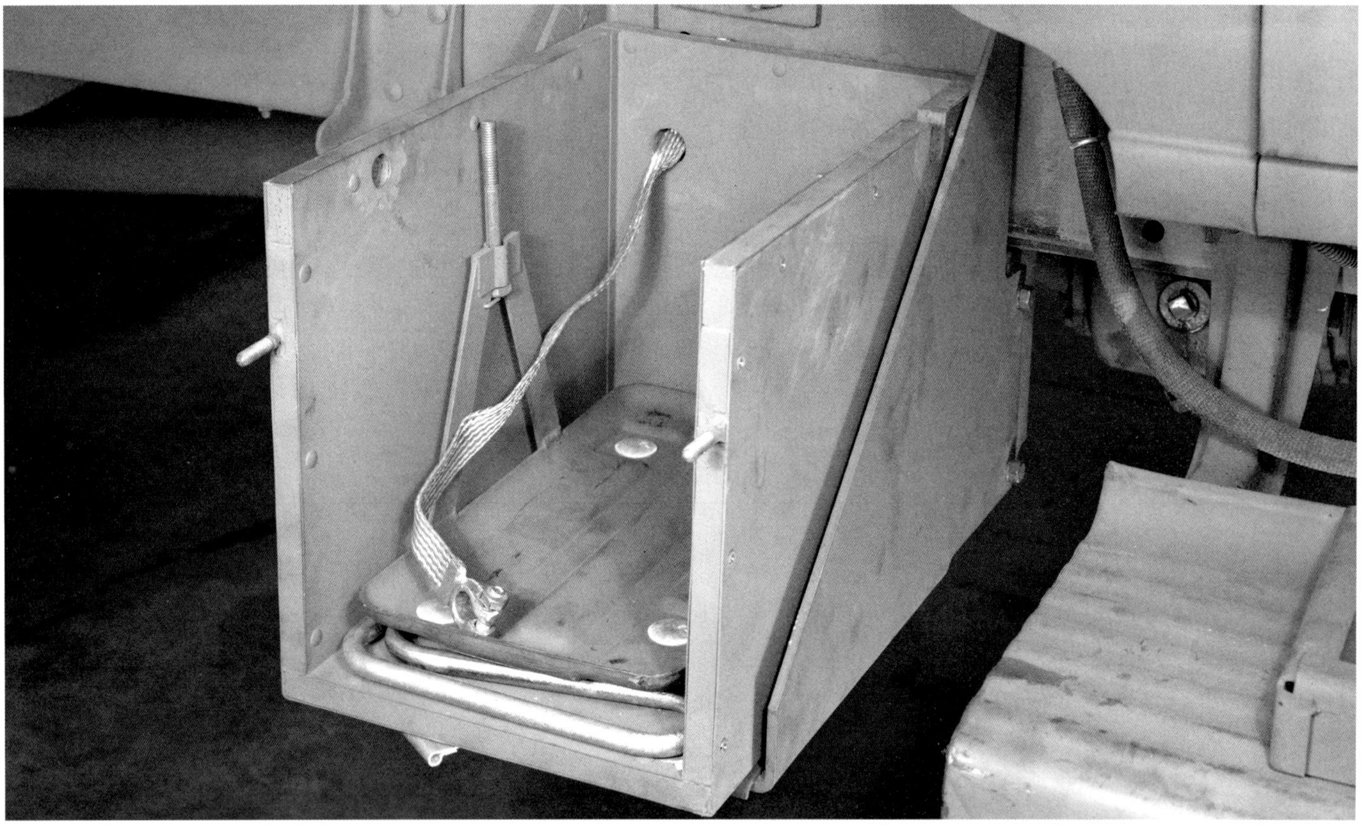

Above: *The battery tray is installed in the winterization battery box. Lying on the tray is the ground strap. The object shaped like an inverted Y on the rear side of the battery tray is the rear hold-down for the battery retainer.* General Motors LLC

Below: *The insulated battery box is shown all buttoned up. Two nuts with single wings secure the box cover in place.* General Motors LLC

Above: *A canvas cover for the radiator was part of the winterization kit for the G-506 Cargo Truck. The cover is shown here with the top half folded down. It was attached to the front of the radiator with snaps. Note the two snap studs toward the top corners of the brush guard.* General Motors LLC

Below: *The radiator cover is fully closed. There were three snaps on each side of the cover and an access flap at its upper center. The technical manual for the winterization kit referred to this cover as "canvas curtain."* General Motors LLC

Above: *The winterization kit included a fuel reserve tank, to be mounted on the right side of the cowl of the G-506 Cargo Truck. This tank provided fuel for the gas-operated heaters. A filler cap was on top of it, and on the bottom was a drain cock and a shutoff valve, with a fuel pipe exiting from the bottom of the tank and entering the cowl.* General Motors LLC

Below: *On the right half of the hood of the winterized G-506 was a small vent for the flue of a gasoline-fueled coolant heater. It had a conical base with a round cap on top, supported by two metal straps.* General Motors LLC

On the right side of the firewall of the bomb service truck are two elements of the winterization kit: at the top is the reserve fuel tank for the under-chassis heater, and below it is the auxiliary battery heater in its stored position. To the right of those two elements is a flexible exhaust duct, attached to the top of a coolant heater at the bottom of the photo. General Motors LLC

The engine compartment with the winterization kit installed is viewed from directly to the right, showing most of the elements depicted in the preceding photo. The winterization kit included numerous hoses, pipes, couplings, and shutoff valves to accommodate the several heating components of the system. General Motors LLC

Above: *A downward view into the right side of the engine compartment of the G-506 vehicle includes the coolant heater and exhaust hose (center) and the auxiliary heater for the battery in its storage bracket (left). Also in view is the distributor.* General Motors LLC

Below: *In a view under a G-506 looking forward, with the oil pan of the engine to the left, at the center is the gasoline heater for the block coolant. Two round covers at the bottom of the heater are open. These covers are linked together with a bar that is connected to a hinge. The drive shaft to the front axle has been disconnected.* General Motors LLC

Above: *The winterization kit incorporated new plumbing on both sides of the engine; this photograph documents the additions on the left side. Components of the engine are seen to good advantage, including the air cleaner, below which are the carburetor and the governor. Note the vehicle oil can stamped "GM" to the left, below the Delco-Remy regulator.* General Motors LLC

Below: *A different design of winterization kit is installed on a G-506 truck. This includes on a side panel of the hood a slave receptacle for external power and a hood latch; a block coolant heater with a flexible exhaust hose connected to a rigid exhaust vent; and the relocated battery.* General Motors LLC

At the center of this photo of the right side of the engine compartment of a G-506 Chevrolet is a black engine oil diluter. In extremely cold areas, the crankcase oil was diluted with a modest amount of gasoline to thin it, so as to produce less resistance to turning over when starting. Once running, the gasoline then evaporates, however care had to be taken to not over dilute, which would bring about improper lubrication and engine damage. Below the diluter is the fuel pump. At the top is the carburetor air cleaner, with the rocker-arm-cover air cleaner below it. General Motors LLC

The same winterization kit shown in the preceding photo is viewed from a perspective more to the rear, with the battery removed. The manner of clamping the exhaust hose and vent to the top and lower right radiator braces is depicted. General Motors LLC

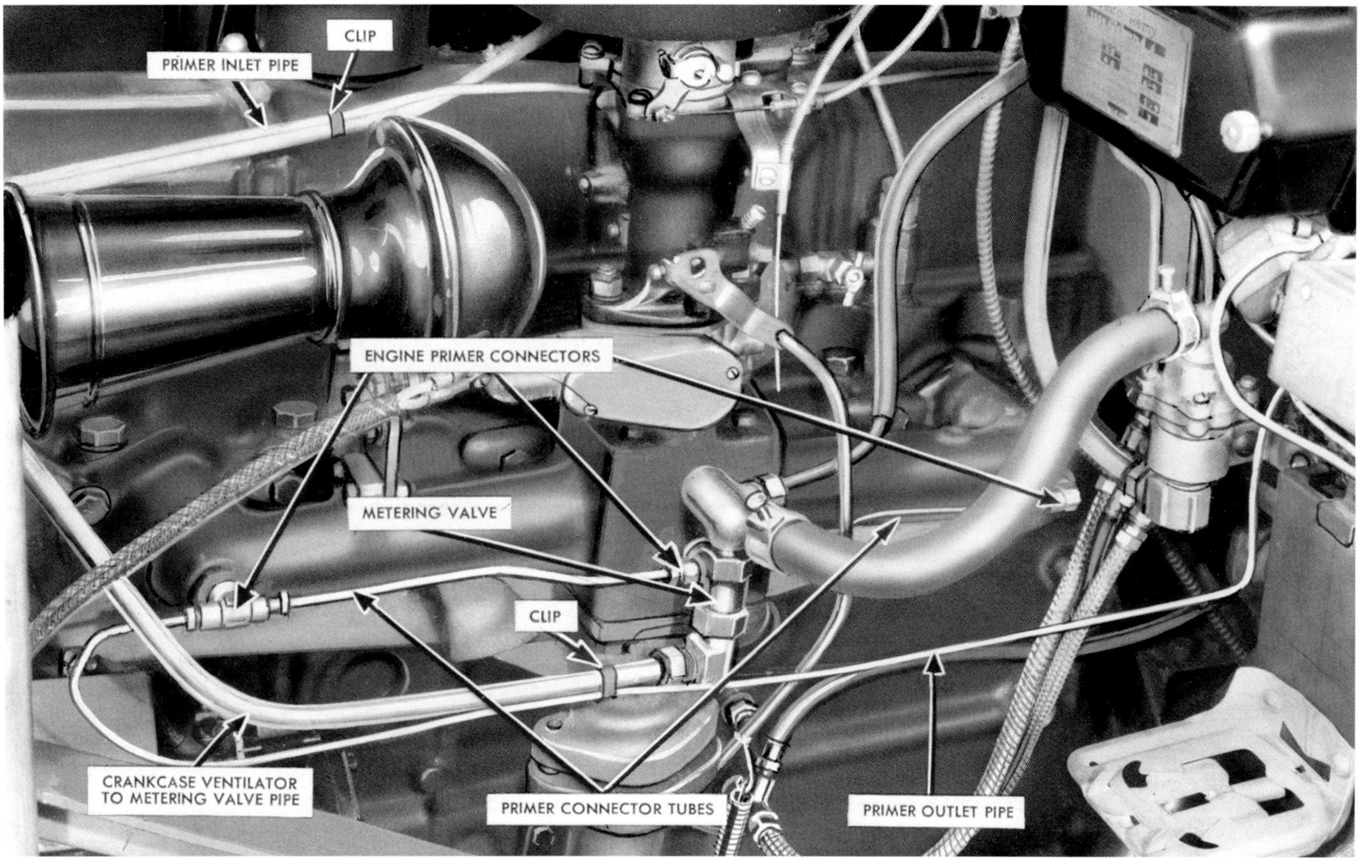

Above: *Further details of plumbing associated with the winterization kit are depicted on the left side of the engine of a G-506 Chevrolet truck. Also in view are the horn at the upper left and the holder for the vehicle oil can at the lower right.* General Motors LLC

Below: *The engine is viewed from directly to the left with winterization components installed. Clear views are available of the carburetor air cleaner, carburetor, governor, intake manifold, exhaust manifold, regulator, and other systems.* General Motors LLC

Above: *In a view inside a G-506 Chevrolet Cargo Truck, two hot-water crew heaters are mounted under the dashboard, part of a winterization kit. Associated hoses and pipes also are in view.* General Motors LLC

Below: *A different design of hot-water crew heater is mounted above the lower part of the steering column in this M6 Bomb Service Truck. Water hoses are routed through the firewall to the heater.* General Motors LLC

Above: *On the Chevrolet assembly line during production of G-506 trucks, axles are being mounted on a chassis frame. The frame is turned upside down, and workers are dropping a rear axle unto the leaf springs. Farther down the assembly line, the chassis will be turned upright, and wheels and tires will be mounted.* General Motors LLC

Below: *Chevy workers are mounting shock absorbers on the banjo-type front axles (foreground) and rear axles (background) of a G-506 truck. The cross-hatched reinforcing ribs on the front differential are visible; to the rear of the front axle, the tie rod is partially visible. The transfer case and drive shafts have been installed.* General Motors LLC

Above: *Chevrolet BV 235-cid six-cylinder inline engines are being packed into wooden crates. These may have been replacement motors, and, judging by the heavy-duty sealing paper lining the crate, the engines are being packed for long-distance, possibly overseas, transport. The worker holding a roll of tape is sealing the ports on the thermostat housings and other openings in the engines. To the left are more engines on roller conveyors.* General Motors LLC

Below: *Toledo Transmission, in Toledo, Ohio, produced transfer cases for GM and Chevrolet during World War II. Here, workers are assembling transfer cases for use in Chevrolet G-506 trucks. Stacked neatly behind this assembly line are gears and shafts. Above the workers in the foreground is an overhead conveyor, consisting of a track with rollers, a chain drive, and racks for hanging parts and components on.* National Archives

Above: *At a plant in Flint, Michigan, assembly-line workers are assembling pintle hooks for military trucks. The man to the left is fitting a coil spring over the shaft of the pintle hook. Springs and hooks are stored on the table in the foreground.* National Archives

Left: *A cab for a Chevrolet 1 1/2-ton 4x4 is being fabricated. The firewall has been installed, and visible inside the cab through the right door is the riser for the right seat. Two "Xs" are stamped in the front of the firewall to reinforce the sheet metal.* General Motors LLC

Above: *In this part of the Chevrolet assembly line, workers are making adjustments to the doors on cabs for G-506 1 1/2-ton 4x4 trucks. In the days before the smaller power hand tools were ubiquitous on automotive assembly lines, the man working on the door of the second truck in line is using a push-twist screwdriver.* General Motors LLC

Below: *Using an overhead crane, workers at Chevrolet's Atlanta, Georgia, plant are lowering an engine and transmission assembly onto a 1 1/2-ton 4x4 chassis. At this point in assembly, the wheels have not been mounted. At this point in production, the frame and springs are painted black, it will be painted lusterless Olive Drab like the axles before the truck is completed.* National Archives

Above: *At the Atlanta plant, workers are positioning a radiator-fender subassembly on a 1 1/2-ton 4x4 chassis. The engine and steering column already have been installed. The bright, thus-far unpainted, plated bolts and lock washers visible on the brush guard were part of the radio-suppression equipment of the truck.* National Archives

Below: *The assembly-line worker to the right is using an air impact wrench to fasten lug nuts on the left rear wheel of a G-506 Chevy truck chassis. Overhead conveyors are delivering wheel assemblies with tires mounted to the wheel-mounting station on the line.* General Motors LLC

The cab-drop station on the Chevrolet assembly line is depicted. Note fenders and bumper are not yet painted OD. Also note the cover on the steering wheel to keep it from being scratched during the body drop. The oil filter has been assembled onto the firewall. The sticker on the windshield says "Special Order Number." General Motors LLC

Chevy workers are using an air wrench to bolt a hood to a cab. This is a very early G-506 Chevrolet, as it has the early-style louvered blackout marker lamps. On this vehicle, the front bumper and fenders appear not yet to have not been painted matte Olive Drab, as these components are of a glossy finish, compared with the matte finish on the other parts of the front end. General Motors LLC

Above: *At the end of the assembly line at Chevrolet's Norwood, Ohio, plant, workers apply the final touches to 1 1/2-ton 4x4 cargo trucks while the man in the dark clothing toward the right goes over a checklist. The body on the truck in the foreground is an early, welded-steel type.* National Archives

Below: *In another photo taken at Chevrolet's Norwood plant, from front to rear, a Chevrolet civilian 4x2 cab-and-box truck, a 1 1/2-ton 4x4 telephone maintenance truck for the U.S. Army, a civilian 4z2 utility chassis and cab, and a 4x2 cab and box for the U.S. Navy are coming off the final-assembly line. The 4x4 truck has the asymmetrical brush guard.* National Archives

Above: *Hundreds of trucks, including many Chevrolet 1 1/2-ton 4x4 trucks, are lined up at an export plant, where they are being held prior to shipment overseas. These vehicles had already been road-tested and approved by Army inspectors, and they were awaiting disassembly and packing in weatherproof boxes for shipment to U.S. Forces overseas, as well as to the Allies: principally the Soviet Union and Great Britain.* General Motors LLC

Below: *Chevrolet 1 1/2-ton 4x4 trucks are parked at a facility where they will be disassembled and packed for shipment overseas. In the row in the foreground are at least three G-506 Chevrolet Truck-Tractors: registration numbers 3106236, 3106212, and 3106227. Several trucks with cargo or dump bodies are in the right background.* General Motors LLC

Above: *At an unidentified Chevrolet plant where military trucks are disassembled and packed for shipment overseas, two Chevrolet 1 1/2-ton 4x4 trucks are parked alongside a stack of shipping crates beneath an overhead crane with the automaker's name on it. The two trucks, registration numbers 386882 and 386864, were listed on the rolls as cargo trucks, although bodies are not mounted.* General Motors LLC

Below: *Three G-506 Chevrolets (the rear end of one is visible to the left) are lined up next to a consignment of crates on a railroad flatcar at an unidentified packing center. Three more G-506 trucks are visible in the right background.* General Motors LLC

Above: *The same load of crates containing disassembled trucks shown in the preceding photograph is more fully visible in this photograph.* General Motors LLC

Below: *A line of eight G-506 Chevrolet vehicles is standing by railroad tracks at a vehicle-packing plant. In the background, workers are securing packing crates on a railroad flatcar.* General Motors LLC

Above: *Many Chevrolet 1 1/2-ton 4x4s were partially disassembled after they were completed and were packed into wooden crates for shipment overseas. In this photograph, workmen are packing axle assemblies into wooden crates, four axles per crate. This plant was the General Motors facility at Linden, New Jersey, as the worker in the right foreground is wearing a union cap marked "Linden Division, Local 595."* General Motors LLC

Left: *At this point in the production line, workers in the foreground are dismantling a truck in order to pack it for overseas shipment. The crew to the right is removing an engine. At the center, a man is pushing a cart full of parts. In the background, cab assemblies are being lifted from chassis.* General Motors LLC

Above: *Workers at a vehicle-packing plant are making adjustments to a disassembled G-506 Chevrolets. Components of two vehicles, including their chassis frames, are carefully arranged on a wooden base, with steel strapping used to secure the components. Some parts are packed in boxes and wrapped with tough paper for weatherproofing. In the foreground of the closest package, the hood of the truck is on top of the chassis frame; note the front leaf springs in the foreground.* General Motors LLC

Below: *In a view dated 1 October 1945, workmen in the garage at the Studebaker Proving Ground, near South Bend, Indiana, are conducting a test whereby they are assembling twin-unit packs of Chevrolet 1 1/2-ton 4x4 Cargo Trucks under field conditions, using no special tools or equipment. The two closest chassis have front winches. After reassembly, workers gave the vehicles a mechanical inspection and then took them on 100 miles of break-in driving.* TACOM LCMC History Office

Above: *Completed Chevrolet 1 1/2-ton 4x4 Cargo Trucks with winch are being driven out of a factory facility. The registration numbers on the vehicle to the front, and the one behind it, W-333465 and W-340527 (or W-340627), pertained to contract W-398-QM-9402, and the Army accepted these vehicles, which were built in Detroit, in 1942.* General Motors LLC

Below: *A variety of trucks, including numerous Chevrolet 1 1/2-ton 4x4 Cargo Trucks, are parked in a holding area at the same plant shown in the preceding photo.* General Motors LLC

Above: *The registration numbers of the Chevrolet 1 1/2-ton 4x4 Cargo Trucks in the first row at an unidentified factory identify them as produced under contract W-398-QM-10328. The nearest four vehicles have, from left to right, registration numbers W-348997, W-349012, W-348963, and W-349007.* General Motors LLC

Right: *Two G-506 Four G-506 Chevrolets are loaded in a boxcar for shipment. Just visible through the door of the boxcar at upper left is the rear of a truck, suspended at an angle above the trucks resting on the car floor, the front ends of which are secured with chains with turnbuckles. Note the spare tire lying on the floor.* General Motors LLC

Chapter 2

Long Wheelbase Cargo Trucks

While the vast majority of the wartime 1½-ton 4x4 Chevrolet trucks were 145-inch models, they were far from the only type produced. In fact, those 145-inch trucks were not even the only cargo trucks produced. In addition, there were 175-inch wheelbase conventional cab trucks with fifteen-foot cargo beds manufactured. These trucks are referred to hereafter as long wheelbase trucks.

Contract W-398-QM-8278

During the initial bid process of July 1940, among the truck types solicited were 90 conventional trucks with 15-foot cargo bodies, without winch. However, as explained in detail in chapter 1, the contract for the construction of these trucks, W-398-QM-8278, included a clause which allowed for the number of vehicles to be increased, and in the case of the trucks with the 15-foot cargo body, the quantity was increased to 219 (218 of which were accepted per the *Summary Report of Acceptances Tank-Automotive Material*). The first 81 of these were accepted in January 1941, followed by 109 in February and the final 28 in March 1941.

The 175-inch wheelbase trucks bore the serial prefix of YQ, as opposed to the YP assigned to all the 145-inch wheelbase trucks on this contract. Also, while the YP trucks had a variety of Master Book Symbols, for the long trucks YQ was both the serial prefix and the Master Book Symbol. The initial order of 90 trucks were

Chevrolet built a long-wheelbase (175-inch) version of the 1 1/2-ton 4x4 Cargo Truck. These vehicles had a 15-foot cargo bed: 6 feet longer than those of the short-wheelbase vehicles. Chevrolet referred to these long-wheelbase trucks by their model number, G-7127. The long-wheelbase trucks are quickly distinguishable by the seven stake pockets on each side of the steel cargo body, compared with five per side on the short-wheelbase trucks. "CHEVROLET" is embossed in the side of the hood to the front of the louvers on this vehicle. National Archives

Above: *A long-wheelbase Chevrolet 1 1/2-ton 4x4 Cargo Truck is viewed from the left side on a snowy day. On these trucks, the rear wheels were positioned below the center of the cargo body. The cab and front clip was the same as on the short-wheelbase vehicles.* Military History Institute

Below: *A long-wheelbase Chevrolet 1 1/2-ton 4x4 truck is observed from the left rear. A full set of mudguards is installed on this vehicle. The spare tire was mounted below the front left of the cargo body, as on the short-wheelbase vehicles.* General Motors LLC

From the rear, the long-wheelbase vehicles were virtually indistinguishable from their short-wheelbase counterparts. The rear curtain of the tarpaulin had a flap that could be raised to form a window. General Motors LLC

A long-wheelbase 1 1/2-ton 4x4 Cargo Box is seen in a Chevrolet factory with the tarpaulin installed, the rear curtain removed, and the tailgate lowered. Note the lashings for the front curtain, twisted around the front bow and stakes. General Motors LLC

A right-side view of a long-wheelbase Chevrolet 1 1/2-ton 4x4 Cargo Truck provides details of the rear drive-shaft assembly. On short-wheelbase trucks, a single drive shaft was employed. On the long-wheelbase trucks, the rear drive-shaft assembly consisted of a forward shaft from the transfer running aft to a support, to the rear of which the forward shaft was attached to a rear shaft running back to the rear axle. To the front of the rear shaft was a universal slip joint. General Motors LLC

assigned US Army registration numbers 328310 through 328400, and the additional trucks on this contract bore registration numbers 331556 through 331683.

Contract W-398-QM-9402

On 23 January 1941, contract W-398-QM-9402 was issued, calling for several thousand additional trucks, including a meager 35 additional 175-inch wheelbase conventional trucks. These trucks included the same improvements as were incorporated in the 145-inch wheelbase trucks on this contract, including the driveline-mounted parking brakes. Thus, these vehicles also got a new Master Book Symbol and serial prefix – ZQ.

Analysis of data in the *"Recapitulation of Facility Expansions for Tracked & Wheeled Vehicles from 1940 to July 15, 1945"* seems to point toward 33 of these trucks being delivered in July 1941 and the final two in August.

W-398-QM-10328 & DA-W-398-QM-8

Two contracts were initially issued for production of the 1942 year model trucks, W-398-QM-10328, which would be renumbered W-374-ORD-2726 following the July 1942 Ordnance Department take over of vehicle procurement, and Defense Aid contract DA-W-398-QM-8. The invitation for bids for both of these contracts had been issued on 6 May 1941, with Chevrolet responding on 14 May. Among the many vehicles on W-398-QM-10328 were a further 20 long wheelbase trucks, while Defense Aid contract DA-W-398-QM-8 included 50 examples for Great Britain.

The contracts to build these trucks were issued on 7 June, and evidence points to these trucks being assembled in December 1941. As these were 1942 year model trucks, with all the upgrades that entailed, the serial number prefix (following the plant code), as well as the Master Book Symbol reflected this. The code assigned was NP.

An additional quantity of 60 long wheelbase trucks was built on contract W-398-QM-10328 in 1942, with 53 being accepted in January and the final 7 in February. With the delivery of those vehicles, production of the long wheelbase G-506 ceased.

From the front, the long-wheelbase Chevrolet 1 1/2-ton 4x4 Cargo Truck appeared very much like the short-wheelbase version. This example had the grille fabricated from steel rods bent accordion fashion. It also has the early-style blackout marker lamps, and no blackout headlight and brush guard extension on the left side. General Motors LLC

Chapter 3

Dump Trucks

An excellent illustration of the challenges of the pre-WWII system of vehicle procurement can be seen in a review of dump truck procurement. Pre-August 1942, dump trucks were procured by the Quartermaster Corps – even though in most instances they would actually be used by the Corps of Engineers. The trucks themselves, however, were built to suit the Quartermasters – not the Engineers.

Engineer Board Report 683 from 20 May 1942, bemoans much about the Chevrolet dump truck, including "The Engineer Board has already recommended the substitution of a 2-1/2 ton cargo truck for the 1-1/2 ton truck as transportation for the footbridge and assault boats, and it would seem desirable at this point to extend the standardization program to eliminate the 1-1/2 ton dump truck. If the recommendations of the

A dump-truck version of the G-506 Chevrolet was produced. The version without a winch went under three successive alphanumerical designations: G-4112 in 1940, G-4152 in 1941, and G-7106 thereafter. The version with a winch was successively designated G-4112, G-4162, and G-7116. Both versions had a 145-inch wheelbase. Dump bodies are sometimes difficult to differentiate from cargo bodies. However, dump bodies did not have the tool box underneath the center rear of the bed and lacked mudguards to the fronts and the rears of the rear wheels (although the mudguards also sometimes were missing on cargo trucks.) Three manufacturers supplied dump bodies and hoists for the G-506: Hercules Steel Products Co., Heil Body Co., and Perfection Body Co. Military History Institute

As seen in a frontal view of a Chevrolet 1 1/2-ton 4x4 Dump Truck with winch, with the dump body raised, there were two guides on the lower front of the body, made of angle irons with curved bottoms, to help keep the body in alignment with the chassis frame when lowered. From this perspective, looking at the raised dump body, the ledges along each side of the body and their beveled fronts are seen to good advantage.
Military History Institute

Above: *A Chevrolet 1 1/2-ton 4x4 Dump Truck without winch is displayed with bows installed on the body. This vehicle is outfitted with troop seats and lazy backs, in conformance with the vehicle's intended multiple use as a dump, cargo, and troop-carrying truck. This is a very-early-production example, with the battery located under the floor of the cab.* National Archives

Below: *In this view of a Chevrolet 1 1/2-ton 4x4 Dump Truck with winch, the release lever for the tailgate is visible at the front of the body. A spare tire is mounted below the front end of the body. The tail pipe is bent up to the rear of the spare tire, with the outlet even with the centerline of that tire.* National Archives

A Chevrolet 1 1/2-ton 4x4 Dump Truck with winch is viewed from the right front with the body raised. No registration number is visible, but this is not a very-early-production example, as it has the frame-mounted battery box and cutaway for the fuel filter. National Archives

other using arms is followed, the 1-1/2 ton dump will be practically the only chassis of the size in the Army. For this reason it should be eliminated."

The report wrapped up with the conclusions including "The 2-1/2 ton 6x6 dump truck should be substituted for the 1-1/2 ton, 4x4 dump truck for future procurement." and recommends:

"a. The 2-1/2 ton 6x6 dump truck be substituted for the 1-1/2 ton 4x4 dump truck in the Tables of Organization for engineer troop units, in the ratio of one to one."

"b. The 1-1/2 ton 4x4 dump trucks now in service be replaced with 2-1/2 ton 6x6 dump trucks as soon as possible."

Contract W-398-QM-8278

Dump trucks were included in the initial contract for production of the 1½-ton 4x4 Chevrolet trucks. That contract, W-398-QM-8278 awarded on 6 August 1940, called for 284 dump trucks with winch and 933 without, and an escalator clause that allowed the government to increase those quantities, within 90 days, by up to 50%.

Deliveries of dump trucks without winch began in November 1940, with six vehicles, followed by 623 more in December. Production of the winch-equipped dump truck began in December, with 41 vehicles delivered. Productions of both types continued into March 1941. By the end of March 1941, 1,558 of the dump trucks less winch had been delivered, as had 284 with winch. This is a clear indication that the escalator clause had been invoked. Like the cargo trucks, the dump trucks on this contract carried the book symbol YP and the Chevrolet model number 4112. The Master Book Symbol for the dump trucks without winch was YH, with YL being assigned to those with winch.

Above: *A dump truck is displayed with the body lowered. In addition to the absence of mud flaps for the rear wheels, visual evidence that this is a dump truck is the tailgate release lever, faintly visible on the front of the dump body.* National Archives

Below: *This dump truck with winch, registration number W-332064, was built under W-398-QM-9402 and was accepted in 1941. It had the battery box on the right side of the chassis frame, with a cutout in the rear corner of the cab to allow clearance for the box.* National Archives

These trucks were equipped with the Hercules Steel Products Company model D12C body, produced in Galion, Ohio. The Hercules literature of the time described their various series of dump bodies in this manner:

"D-12 - A rugged braced side body with wide rub rails, closed side braces, and heavy boxed corners. Heavy duty construction for heavy duty work."

"D-4 - A strong, straight side body for general use. Full length, pressed steel rear and gussets. On bodies of large capacity, tailgate is additionally reinforced...."

"D-2 - Removable side, rub rail, convertible body. Hinged rear corner posts. Converts into platform with tailgate lowered. Sides may be hinged at additional cost."

The D12C beds installed on the YP series trucks certainly seem to match the description above.

The next contract issued, Contract W-398-QM-9402, called for 1,884 more dump trucks, 500 of which were to be equipped with a winch. These trucks, like their predecessors, were 1941 year-model trucks, however now they bore the serial number prefix of ZP, as a result of the previously described improvements to the chassis. A later supplement added 30 more trucks to this contract, but they were 1942 models with a NH serial number prefix. The reason for this supplement no doubt was due to thirty of the ZH chassis (serial prefix ZP) on Contract W-398-QM-9402 being used for crash trucks.

The front winch and the front clip of the same dump truck shown in the preceding photo, registration number W-332064, are viewed close-up. The winch, as with the one on the cargo-truck version, was a Gar Wood or Heil Model 2-U-512, the two types of winches having interchangeable parts. The winch was mounted on an extension of the chassis frame and was powered by a drive shaft from the power takeoff. General Motors LLC

From this frontal perspective, the Chevrolet 1 1/2-ton 4x4 Dump Truck was indistinguishable from its cargo-truck counterpart. This early example with winch, registration number W-332064, has the symmetrical "snake" brush guard and the early, louvered blackout marker lamps. General Motors LLC

A key visual feature of the Chevrolet 1 1/2-ton 4x4 Dump Trucks is evident in this photograph of vehicle registration number W-332064: the cross member below the tailgate containing a recessed tool box, present on the cargo trucks, has been deleted. This was to give free play to the rear of the dump body when it was being hoisted. The two hinges of the dump body are visible below the body. General Motors LLC

Above: *The right midsection of a Chevrolet 1 1/2-ton 4x4 Dump Truck is viewed close-up. Very new lashing ropes are installed on the tarpaulin. Also present are a reflector on the front of the dump body, the fuel filler between the cab and the body, and the battery box.* General Motors LLC

Below: *At a Chevrolet factory, workmen are lowering a dump body onto a Chevrolet 1 1/2-ton 4x4 chassis without winch. The man to the right is grasping one of the long mounting bolts on the bottom of the hydraulic hoist.* General Motors LLC

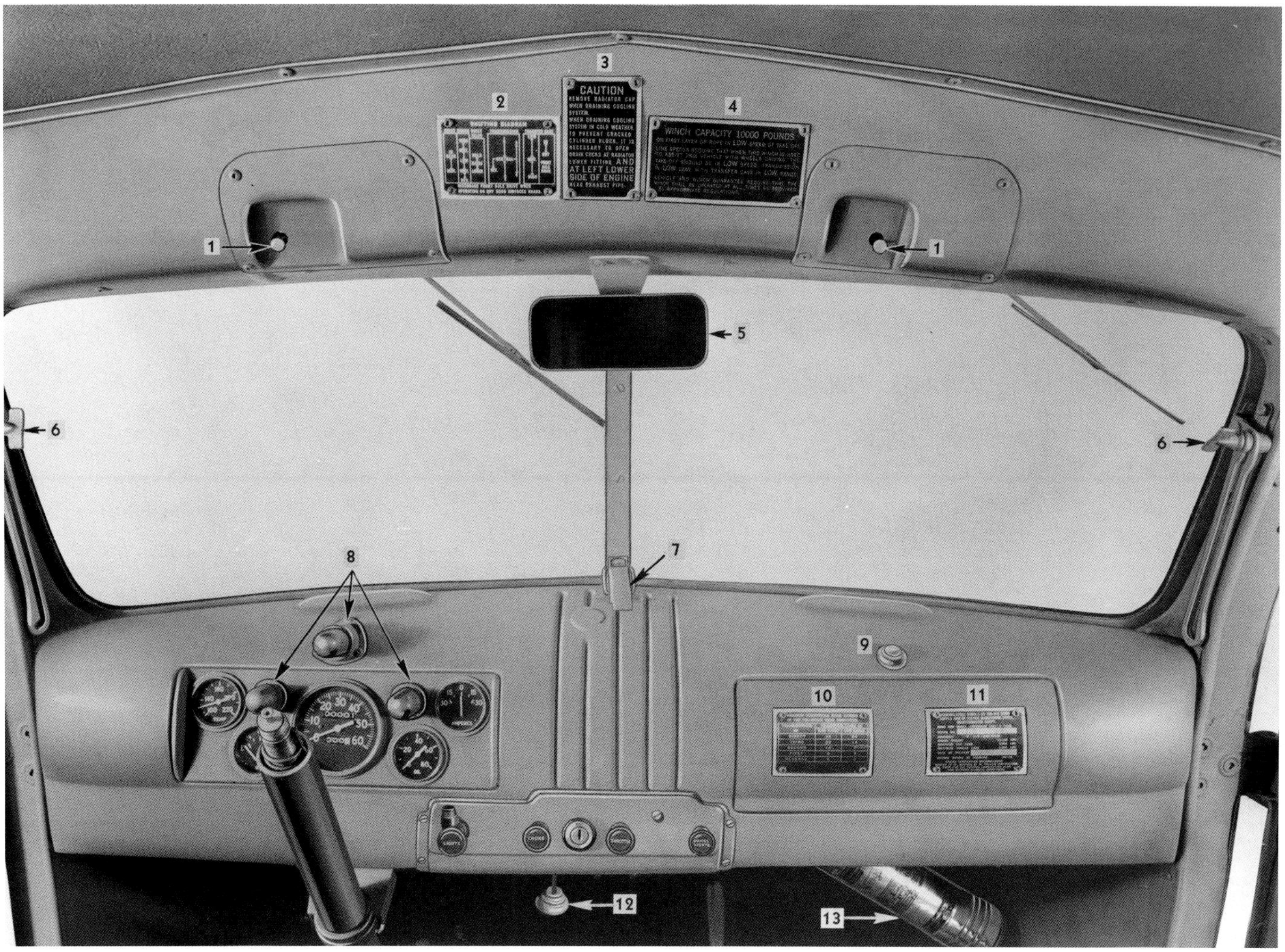

The upper part of the cab of a Chevrolet 1 1/2-ton 4x4 Dump Truck is shown. Data plates are on the front of the ceiling and on the right side of the dash. This vehicle has a military instrument panel. Below the right side of the dash is a fire extinguisher. Hold-open brackets are on each side of the windshield. General Motors LLC

These trucks also used dump beds produced by Hercules, although these trucks did not use the D12C bed, but rather a different model, as shown in TM 10-1430, of June 15, 1941.

In September 1941, Chevrolet began producing the 1942 model year trucks. The initial production of these vehicles were on contracts W-398-QM-10328, which would be renumbered W-374-ORD-2726 following the July 1942 Ordnance Department take over of vehicle procurement, and Defense Aid contract DA-W-398-QM-8. The latter included 350 dump trucks with winch, while the former included 1,150 dumps without winch and 250 with winch. These trucks used Master Book symbols and serial number prefixes of NH for the non-winch trucks, and NL for the winch models.

These trucks were initially equipped with Hercules D12S2 beds, except for the first 31 vehicles, Registration W-352990 through W-353020, which used the same bed and hoist mechanisms as did the ZH trucks on the previous contract.

This information is recorded in the June 1945 ORD-9 which reads, "The first 31 Dump Body Models on Contract W-298-QM-10328 (Registration W-352990 through W-353020) used the same Dump Body Hoist and Power Take-Off Parts as were used on the previous Contract. Accordingly, any parts required to service the Bodies on Parts on Vehicles bearing those Registration Numbers should be selected by referring to Book Symbol ZH, which covers parts for these 31 jobs as well as the previous contract."

"For NH trucks with registration numbers W-352990 through W-353020 inclusive, use dump body power take off parts designated by model symbols ZH."

CONTROLS AND INSTRUMENTS, 4x4 TRUCKS

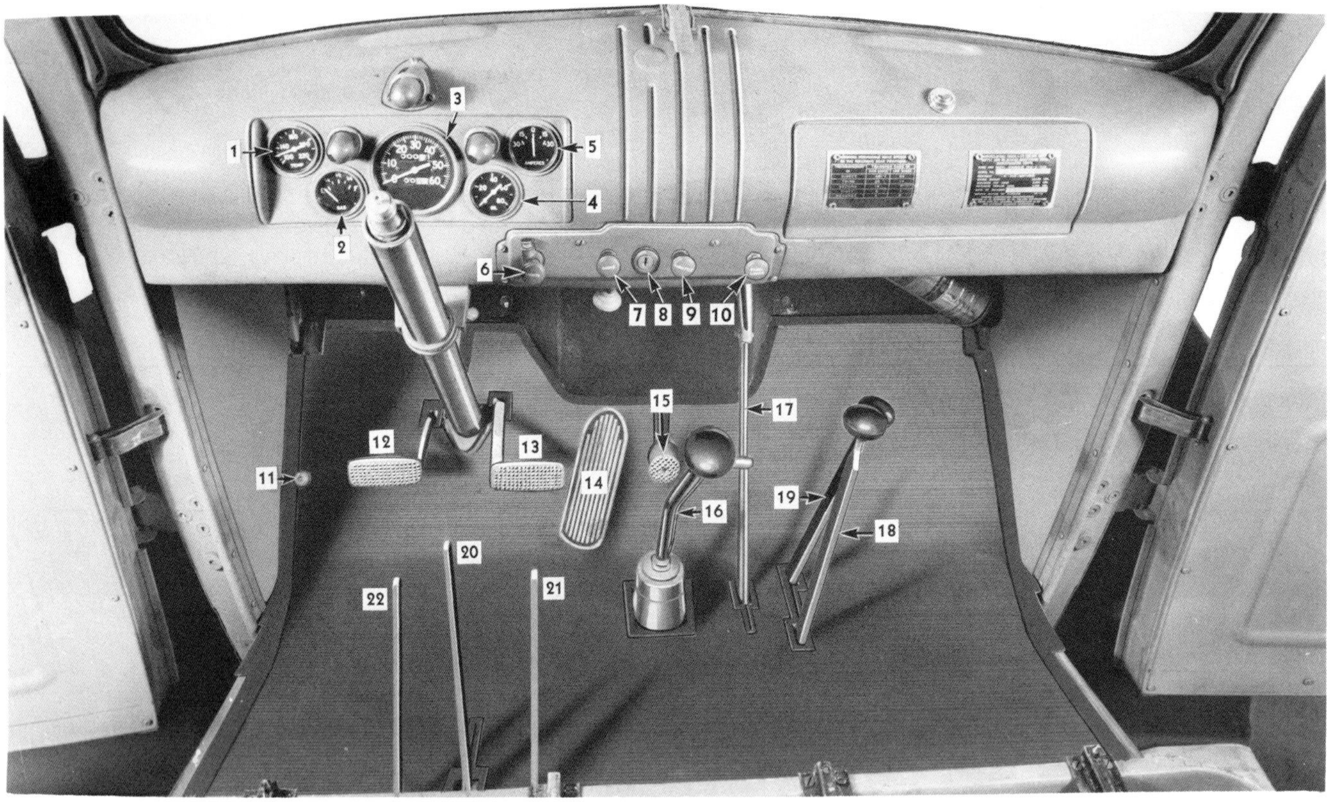

1—Temperature Indicator
2—Fuel Gauge
3—Speedometer
4—Oil Pressure Gauge
5—Ammeter
6—Lighting Switch
7—Carburetor Choke
8—Ignition Switch
9—Hand Throttle
10—Instrument Light Switch
11—Headlight Dimmer Switch
12—Clutch Pedal
13—Brake Pedal
14—Foot Accelerator
15—Starting Switch Pedal
16—Transmission Gearshift Lever
17—Hand Brake Lever
18—Transfer Case Shifting Lever
19—Front Axle Control Lever
20—Winch Power Take-off Control Lever
21—Hoist Valve Control Lever
22—Hoist Power Take-off Control Lever

Chevrolet Motor Division
Detroit, Michigan

Elements on the lower part of the front of the cab are displayed. The details are similar to those of the G-506 cargo trucks, except for the three levers at the lower left. These controlled, from left to right, the hoist power takeoff, the winch power takeoff, and the hoist valve. General Motors LLC

The prior contract referred to is of course Contract W-398-QM-9402.

The Hercules body used on the remainder of the trucks on contracts W-398-QM-10328 and DA-W-398-QM-8 was the company's D12S2 bed. This bed was of noticeably lighter construction that the D-12C used on the first contract. No doubt this reflected the desire to conserve steel, which had brought about the introduction of wooden cargo beds as well.

Ultimately, a total of 1,150 dump trucks without winch, and 250 dump trucks with winch, were ordered on contract W-398-QM-10328. During 1941 642 of the trucks without winch and 129 of the trucks with winch were delivered, with the remainder of the trucks on this order delivered in 1942.

With regard to the Defense Aid contract DA-W-398-QM-8, the Chevrolet Master Parts Price List 1941-1942 (TM 10-1475), issued 1 April 1942, states that 350 trucks were produced. This number is echoed by "Ordnance Department Administrative and "Tactical Vehicles, 1940-145 Defense Aid," dated 22 May 1945, published by the Office Chief of Ordnance – Detroit. Later, four additional dump trucks with winch were purchased on contract W-398-QM-11510.

However, British records indicate that they only received 331 such vehicles, which were given British registration numbers L777093 through L777366 and L777384 through L777440. "Tactical Vehicles, 1940-145 Defense Aid" reported that these trucks are all G-7116NL, as would be expected.

The demand for dump trucks and beds was such that Hercules could not alone supply Chevrolet. Tapped to supplement the Hercules production was Heil, who

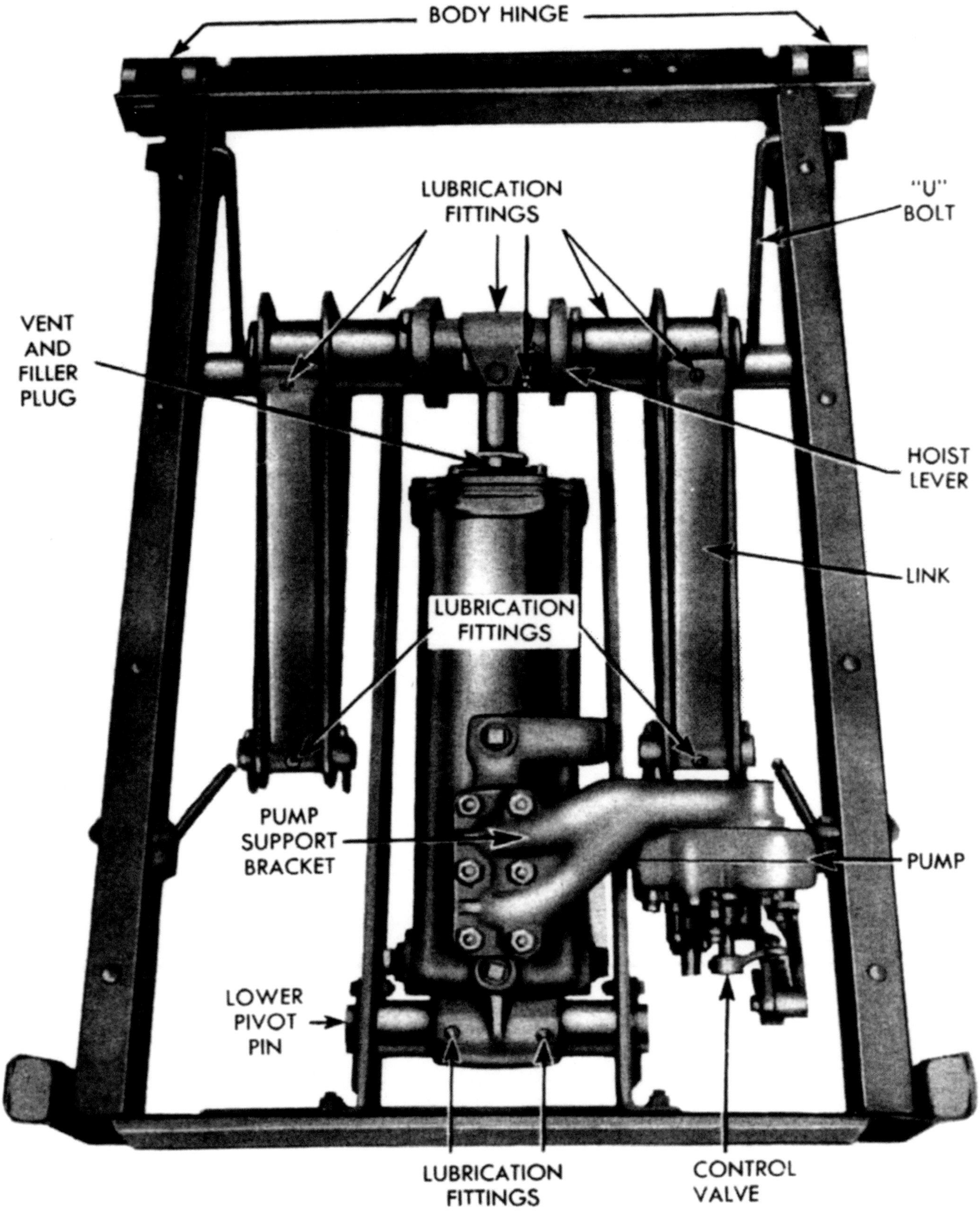

Hydraulic hoists by three different manufacturers were employed on the Chevrolet 1 1/2-ton 4x4 Dump Trucks. The example shown here was the Heil SL-1 Hydraulic Hoist. A key feature for identifying this hoist was the Y-shaped hydraulic manifold, connecting from the pump to the cylinder.

supplied their model 367-6-DA dump bed, using the SL-1 hoist.

By December 1943, yet another dump body manufacturer had become involved, the Perfection Steel Body Company, who supplied their 7313 body and 615 Special hoist.

The next group of dump trucks built were 500 trucks with winch on Quartermaster contract W-398-QM-10939 (which was later known as Ordnance contract W-374-ORD-2733) which were contracted for in October 1941 and delivered in 1942. These trucks, which had Master Book Symbol and serial number prefix NL, bore Army registration numbers W-362127 through W-362626.

By far the largest contract for the production of dump trucks was W-398-QM-11510. In addition to the four winch-equipped trucks built for the British on this contract, 3,233 trucks were built for the US. Of these,

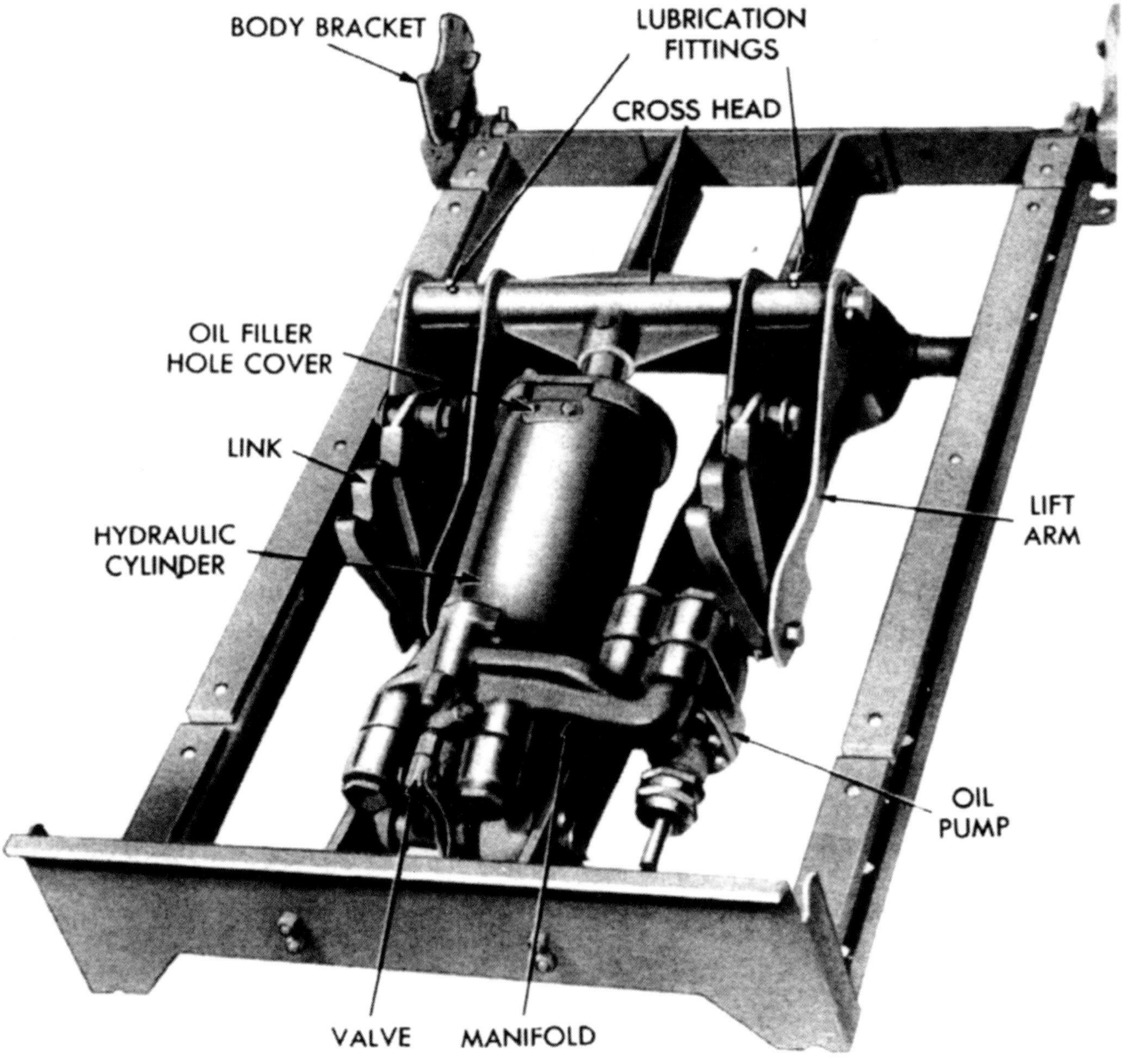

The Hercules K.D.O. Hydraulic Hoist also was used on the Chevrolet 1 1/2-ton 4x4 Dump Trucks. The designs of the hoist subframes and body hinges varied from maker to maker. The driver controlled the operation of the hoist from the cab.

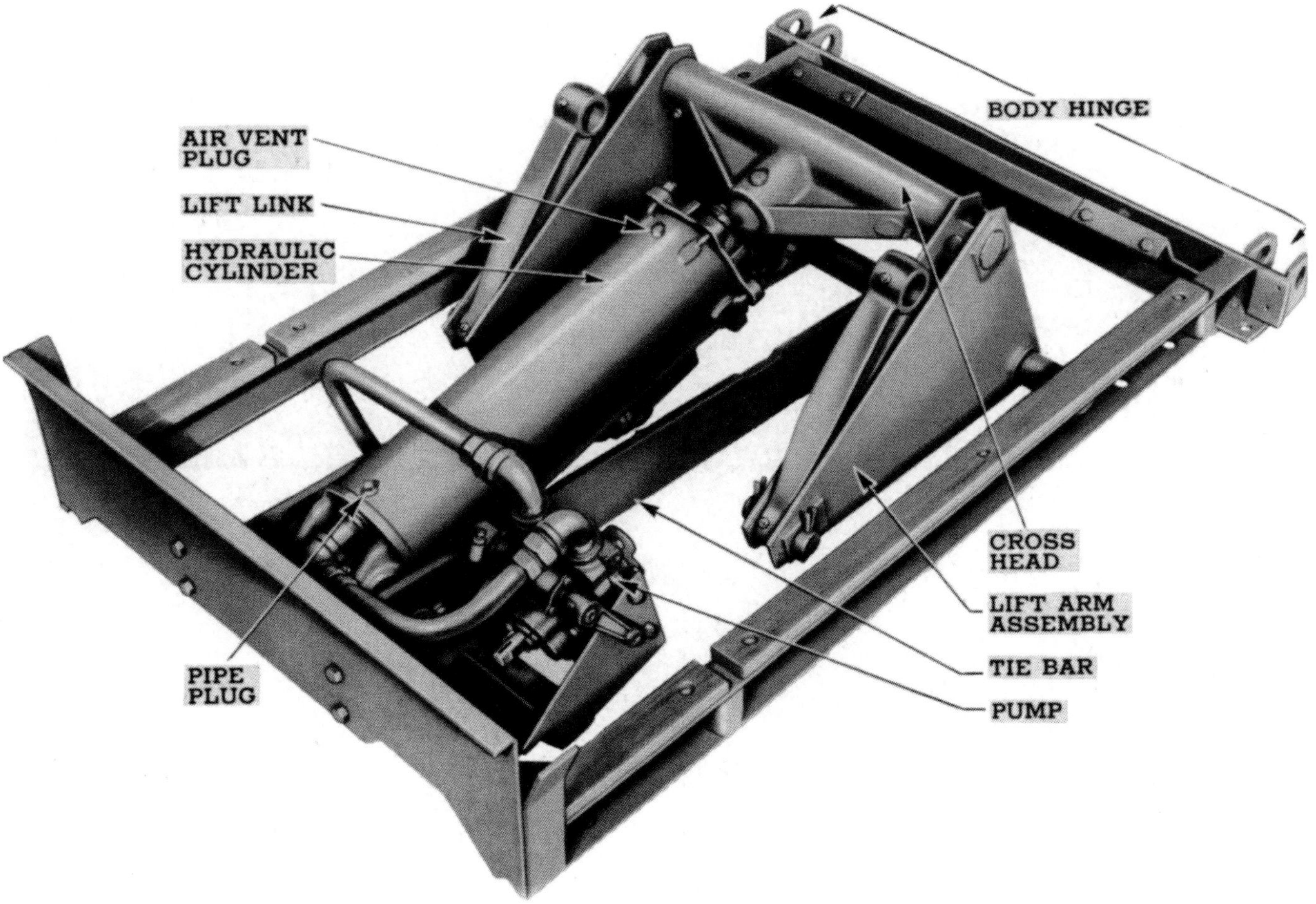

The third hydraulic hoist was the Perfection 615 Special. Rather than having a prominent hydraulic manifold on top of the rear of the cylinder, the Perfection 625 Special had exposed hydraulic lines.

3,198 were built in 1942, with 35 more in January of the following year. These were joined by 4,886 dump trucks without winch built on the same contract and extensions. Production of the non-winch dump truck continued until May 1943, when the final 8 were delivered.

When those last 8 dump trucks were delivered in May 1943, it was anticipated that they would be the last dump trucks on the 1½-ton chassis, owing to the deficiencies pointed out in the May 1942 Engineer Board report cited earlier in this chapter.

Nevertheless, the Russians requested 104 more dump trucks without winch, which were built in January 1945. To date no documentation has surfaced detailing the specific type of Hercules bed which was installed on these trucks, although they were procured on contract 33-019-4944.

Ultimately, 9,008 dump trucks without winch, and 5,133 dump trucks with winch, were accepted by the US Government, either for their own use, or on the behalf of Defense Aid recipients.

Above: *This Chevrolet 1 1/2-ton 4x4 Dump Truck was one of the vehicles sent to the British under Lend-Lease. The tarpaulin has a two-color camouflage pattern, as was common practice with the British. Marked in chalk on the door is "S/M 2229" over "CHEV." "S/M 2229" (Supply / Mechanical 2229) was the British contract under which this vehicle was procured: this contract included registration numbers 777093 to 777366.* Imperial War Museum

Below: *Chevrolet 1 1/2-ton 4x4 Dump Truck registration number W-337248 was accepted in 1941 and was produced under contract W-398-WM-9402. The right taillight is visible alongside the chassis frame to the rear of the rear tires.* Keith Buckley collection

Above: *The registration number on this Chevrolet 1 1/2-ton 4x4 Dump Truck, 3100950, identifies it as a 1942-model vehicle manufactured under either contract W-398-QM-11510 or W-374-ORD-2694. Trucks built under these contracts were produced at the Atlanta, Baltimore, Norwood, Oakland, St. Louis, and North Tarrytown plants. Note the other G-506 truck, without a body, to the right.* General Motors LLC

Below: *A Chevrolet 1 1/2-ton 4x4 Dump Truck with winch, with no registration number on the hood, is viewed from the right side. This vehicle has the welded-steel dump body and frame mounted battery box.* General Motors LLC

Chapter 4

Telephone Line Maintenance and Telephone Body with Earth Borer

In today's world of radios, cell phones and satellite communication, it is easy to forget just how important wire communications were during the WWII era. Whether voice over wire or telegraph, wire communication carried the bulk of army communications.

In his famous speech to the Third Army, General George S. Patton, Jr gave an indication of just how important this work was, "One of the bravest men I saw in the African campaign was on a telegraph pole in the midst of furious fire while we were moving toward Tunis. I stopped and asked him what the hell he was doing up there. He answered, 'Fixing the wire, sir.' 'Isn't it a little unhealthy up there right now?' I asked. 'Yes sir, but this damn wire has got to be fixed.' I asked, 'Don't those planes strafing the road bother you?' And he answered, 'No sir, but you sure as hell do.'"

Accordingly, three special truck models were produced for the Signal Corps for use in telephone wire line installation and maintenance. These three models were:

- Line Maintenance no winch K-42 (LC-102)
- Line Maintenance K-43 (LCW-102)
- Earth Boring and Pole Setting K-44

Two of the three models were equipped with a winch, and one was not. But, these are not winches normally associated with the 1½-ton 4x4s, which were mounted behind the front bumper. Rather, these winches were mounted at the front of the bed, behind the cab.

At the time these trucks were being developed, American Telephone and Telegraph (AT&T) was the world leader in telephone communication, and their Bell System spanned the nation. The Signal Corps wisely opted to pattern the Army's wire telephone service on that of "Ma Bell," the truck bodies, tool and components, and even the storage layout of these items were patterned on the Bell System. The wisdom of this would pay huge dividends during the war, as thousands of Bell linemen volunteered and were drafted. These soldier telephone men went overseas with equipment that they were already intimately familiar with.

Work on these special trucks for the Signal Corps began in late 1940, making them one of the earliest special variants of the 1½-ton Chevrolet considered. Pilot models of the telephone construction truck and earth boring truck inspection was held at a GM facility in Detroit on December 26, 1940. However, once production began, the truck chassis for these vehicles were shipped by rail to Cleveland for body installation by American Coach and Body Company. American, an industry leader in the telephone and utility body field, occupied a manufacturing facility that was previously the Peerless Motor Company.

October 1940 letter from the Office of the Quartermaster General to the Chief Signal Officer, ……."Invitation for Bids 398-41-75, Trucks 1 ½-ton (LC), 4x4, Telephone Construction and Maintenance, requested in letters of your office dated August 15 and September 18, 1940, File: R-8007." 'LC' indicates Light Chassis.

The K-42

Surviving documentation states that the Signal Corps had planned to roster a total 299 of the line construction trucks without winch.

However, on 5 August 1941 the Signal Corps Board issued a report on case number 402, in response to a directive from the Chief Signal Officer issued on 21 May 1941, evaluating the then-new trucks.

With regard to the K-42, referred to by its American Coach and Body Company designation LC-102, the report states "...further procurement of this type of vehicle is not warranted. In 1938, when the Signal Corps Board recommended the issue of Trucks LC-102 (equivalent of LC-104 and LC-109, except for body length to signal battalions, it was contemplated that this particular vehicle would be utilized as an installation truck by the operation company of the battalion and not by the construction company for line construction purposes. Because this vehicle is not equipped with a winch and derrick, it is practically useless as a line construction

In addition to the Chevrolet Pole-setter Truck with Earth Borer K-44, the U.S. Army Signal Corps fielded another specialized truck based on the Chevrolet 1 1/2-ton 4x4 chassis: the Telephone Maintenance Truck K-42. This vehicle's body had its origins in the Bell Telephone KC-series bodies and had an entirely different body from that of the K-44, also lacking the center winch, earth borer, and derrick of the K-44. Seen in this photograph is K-42 registration number W-003026, a Detroit-built, 1941-model-year vehicle produced under contract W-398-QM-8778. The lack of the prominent winch-shaft-extension fitting for the power reel CR on the right side of the body is visual evidence that this vehicle lacked the center winch, and thus was a K-42, and not its sister vehicle, the K-43, which would be introduced subsequently. National Archives

vehicle. Furthermore, the vehicle is not well suited for substation installation operations at an Army or Army Corps Headquarters because the side compartments of the truck are not well adapted for the accommodating small parts associated with subscribers' station installation work and because the truck is far heavier than is necessary for this type of work."

Whether as a result of the scathing review above, or other circumstances, both Chevrolet and Ordnance documents agree only 67 of the K-42 Line Maintenance trucks without winch were accepted by the Army. This arguably makes these trucks the scarcest variant of 1½-ton 4x4 Chevrolet. The US Army registration numbers assigned to these trucks were W-002968 through W-003034.

Fortunately, enough records of these trucks survive to allow us to make some observations about Chevrolet 1½-ton 4x4 production in general.

What is interesting to observe is that counter to what many would believe, the registration numbers were

Registration Number	Chevrolet Serial Number	Date of Delivery	Assignment
W-002968	*		
W-002969	9YP01-4877	1/22/41	50th Signal Battalion
W-002973	9YP01-4953	1/23/41	50th Signal Battalion
W-002974	9YP01-4997	1/24/41	50th Signal Battalion
W-002975	*		
W-002976	*		
W-002977	9YP01-5170	2/4/42	50th Signal Battalion
W-002978	9YP01-5082	1/31/41	50th Signal Battalion
W-002979	9YP01-5134	1/31/41	50th Signal Battalion
W-002982	9YP02-5392	2/6/41	60th Signal Battalion
W-002983	*		
W-002985	9YP02-5436	2/10/41	50th Signal Battalion
W-002986	*		
W-002987	9YP02-5525	2/11/41	50th Signal Battalion
W-002988	9YP02-5522	2/11/41	60th Signal Battalion
W-002989	9YP02-5560	2/11/41	60th Signal Battalion
W-002990	9YP02-5563	2/11/41	60th Signal Battalion
W-002992	9YP02-5612	2/13/41	60th Signal Battalion
W-002994	9YP02-5603	2/14/41	60th Signal Battalion
W-002998	9YP02-5837	2/21/41	60th Signal Battalion
W-002999	9YP02-5892	2/27/41	60th Signal Battalion
W-003000	*		

* These trucks, which were shipped to the Lexington Signal Depot for further distribution, were not as thoroughly documented as the others.

NOT assigned in the same progression as the Chevrolet serial number. Thus, it is impossible to project what any particular trucks serial or registration number may have been based on another truck (or group of trucks) serial or registration numbers.

Not surprising, given the information above, all the K-42 trucks were Chevrolet Master Book Symbol YB trucks with serial number prefixes YP. These chassis for all of these trucks were produced in the Norwood plant. These vehicles were equipped with bodies made by American Coach and Body Company of Cleveland. The American part number for K-42 body was 5061-11. The body for the initial production K-43, discussed in the next section, was American part number 5061-10. Both visually and from the manufacturing standpoint, the bodies were very closely related.

K-43

The K-43 was essentially the same truck as the K-42, with the addition of a winch mounted in the front of the bed, additional power train components to drive the winch, and the A-frame derrick, collapsible cable reel and other gear to use in conjunction with the winch.

When production of the K-43 began, Chevrolet assigned Master Book Symbol YA to the trucks along with serial number prefixes YP. The bodies installed were produced by American Coach and Body Company, and were that company's part number 5061-10. The 1942 Chevrolet Master Parts Price List, TM 10-1475, shows 114 of the Telephone Maintenance Trucks with Winch has having been produced. The US Army registration numbers assigned to these trucks were W-002854 through W-002967. These trucks, along with the 67 K-42

Above: *Although the K-42s lacked the center-mounted winch and detachable rear derrick featured on the K-43 Telephone Maintenance and Construction Truck, the K-42s still had the mountings on the rear of the body for the two rear gin poles of the derrick, and also had in the left side of the body the storage pocket for the gin poles, as seen in a rear view of Chevrolet K-42 W-003026. The spare tire and the jack were stored in the right side of the body. A grab handle was on each side of the body above the mounting brackets for the derrick.* National Archives

Below: *The K-42 was equipped with plenty of storage compartments on the side of the body. The doors to those compartments are open on the right side of K-43 W-003026. Like the early K-44s, the K-42s had the fuel tank between the cab and the body. The body was fitted with a telescoping, sliding roof, shown open in this photo. It was designed to be stored out of the way, as seen here, or pulled back to cover the body.* National Archives

trucks, were built from March 1941 through June of the same year.

Once again, a handful of surviving documents record the relationship between Chevrolet serial numbers, US Army registration numbers and date of delivery to illustrate that these numbers do not increment in the same manner relative to each other. A list of this information as it pertains to some K-43 trucks is shown below:

Unlike the K-42, production of the K-43 line maintenance truck with winch was long-lived, lasting until June 1945. While none were built on the briefly-produced "Z" series chassis, production began in earnest on the "N" series chassis, the trucks being assigned the Master Book Symbol NS trucks, with serial number prefixes NS as well.

The line maintenance bodies of the NS trucks differed from that of the Y-series trucks. Still made by American Coach and Body, the part number was now 5201-25. This body was notably longer than the previous body. With the prior production vehicles, the fuel tank is visible in the space between the cab and the bed – this is not the case with the NS trucks.

The matter of the size of the bed was touched on in the previously mentioned 5 August 1941 Signal Corps Board report. This report states, "Under a recently placed contract, Contract No. W-297-QM-6778, the manufacturer provided a chassis having a wheel based of approximately 145", which chassis, when equipped with a 102" body, resulted in a 10" space between the rear of the cab and the front panel of the body. The specification has been revised to insure the proper relationship between the rear of the cab and the front panel of the special body."

"Provision is made for mounting the 35-gallon gasoline tank in the left front corner of the special body in the event that a tank of this capacity cannot be accommodated on the side of the chassis, in order to obviate the possibility of the manufacturer's mounting the tank between the rear of the cab and the front panel of the body, and thereby attempting to justify a 10" dimension between the rear of the cab and the front panel of the body."

The report also includes some interesting observations concerning these trucks. The interior of the stowage compartments, which were painted in an aluminum paint per AT&T practice (in order to aid visibility in locating small parts), would be painted gray instead moving ahead. This was because even in August 1941, aluminum paint was a strategically critical item and thus in short supply. It was felt that because these compartments

Registration Number	Chevrolet Serial Number	Date of Delivery	Assignment
W-002854	9YP01-4861	4/11/41	60th Signal Battalion
W-002857	9YP01-4258	4/9/41	60th Signal Battalion
W-002863	9YP01-2938	1/23/41	50th Signal Battalion
W-002868	9YP01-4985	4/28/41	60th Signal Battalion
W-002870	9YP01-5036	1/22/41	60th Signal Battalion
W-002871	9YP01-5028	2/8/41	50th Signal Battalion
W-002875	9YP01-5177	1/31/41	50th Signal Battalion
W-002895	9YP02-5898	2/20/41	50th Signal Battalion
W-002901	9YP02-6111	1/22/41	50th Signal Battalion
W-002902	9YP02-6122	2/27/41	50th Signal Battalion
W-002903	9YP02-6139	2/22/41	50th Signal Battalion
W-002904	9YP02-6115	2/27/41	50th Signal Battalion
W-002916	9YP02-6263	4/24/41	60th Signal Battalion
W-002922	9YP09-6432	4/3/41	60th Signal Battalion
W-002928	9YP03-6854	4/9/41	60th Signal Battalion
W-002931	9YP02-6258	4/22/41	60th Signal Battalion

Above: *The Chevrolet K-43 Telephone Maintenance and Construction Truck was similar to the K-42, except that it was equipped with a center winch and derrick, for erecting telephone poles. When not in use, the three gin poles that made up the derrick were stored in a pocket on the left side of the body. When these poles were stored for travel, the driver was unable to enter or exit the cab through the left door. The K-43 body had the same telescoping roof as the K-42.* General Motors LLC

Below: *This retouched photograph with the background whited-out shows a winch-equipped K-43 from the left side. The early K-43s had a noticeable gap between the cab and the body, with the fuel tank located in this gap. The two gin poles that constituted the rear of the tripod derrick when erected are on the bottom of the stack, with the telescoping forward gin pole on top. The forward pole had a semi-enclosed sheave on the top.* General Motors LLC

A sliding window is on the front of the body of the K-43. The storage compartment doors are open; in the forward compartment is a heavy-duty tow chain and several bags, each of which contains a pair of Weed Extra Heavy Truck and Bus Tire Chains. General Motors LLC

were normally closed, the gray color would not interfere with vehicle camouflage. Secondly, the bodies, which previously of painted, but galvanized construction would henceforth be mild steel, as galvanized material was in short supply.

Another easily noticeable difference between the 5061-10 body and the 5201-25 had to do with the sliding cover over the bed differing, and the arrangement of compartment doors on the driver's side was changed as well.

Beyond the body, the cab was revised as well, with the roof of the cab now being punctured by a Cab-O-Scope. This device allowed the driver to see above and to the rear of the truck, especially important when the derrick was being used. The Cab-O-Scope was a large mirror supported by a ball and socket bracket, in turn installed in a housing protruding through the roof of the truck. The Cab-O-Scope presented the driver with a view of all objects in the rear in proper relation to each other and all movements appear in their proper directions. The Cab-O-Scope had been used previously on the Earth Boring trucks, and it was felt worthwhile to include it in the K-43 as well.

Production of the revised NS series trucks began in June 1942. Contract W-398-QM-11508 called for 181 of the vehicles, with a later addition of 26. These trucks were assigned registration numbers W-007213 through W-007293 and W-07932 through W-007957.

When contract W-398-QM-11510 was issued, this called for a whopping 2,463 more examples of the K-43. Thirty-seven of these were delivered in 1942, nine in October, 26 in November and 2 in December. These were followed by 1,967 beginning in February 1943 and continuing through the end of the year, and the last 459 in 1944. Production in December 1942, and for all of 1943 but for September, November and December fell short of planning owing to a shortage of seamless tubing used in the construction of the bed, and the Navy taking over the capacity of the plant that was producing the gears for the rear winches.

In a rear view of a Chevrolet K-43, a tube with a sheave for use with the winch cable is mounted on the brackets at the rear of the body where the rear gin poles of the derrick were installed when in operation. Presumably these brackets served as the mount for the pole and sheave during travel. The brackets where this pole and sheave were installed during winching operations were toward the top of the rear of the body, directly above the two gin pole brackets. Stored on the left side of the interior of the body is the power reel CR. When in operation, this reel was mounted on an extension of the winch shaft on the right side of the body. The winch was located in the enclosure below the window at the front of the body. General Motors LLC

The telescoping roof is extended to cover the entire body in this left-rear photo of a K-43. On the upper part of the left side of the body were holders for an extension ladder: three lower rests and two upper holders. Below the rear of the body are two telescoping supports, to be deployed under certain conditions when the derrick was lifting a heavy load. The supports are in the travel position, secured with chains. General Motors LLC

Defense Aid contract DA-W-398-QM-517 (later known as W-374-ORD-2839) called for a number of the K-43 trucks. The first group of 842 trucks were built in 1944 and were assigned US Army registration numbers 0080014 through 0080855. These were followed by an additional 535 of the vehicles. Thirty-six of those, registration numbers 0097813 through 0098184 were built in 1944, and the remaining 489, registration numbers 0097813 through 0098184, were built from January through June 1945.

However, Lend-Lease documents only record that 60 K-43 trucks were supplied to the British Empire, and 73 of the vehicles were provided to the Free French forces.

Like many special-purpose military vehicles, the basic K-43 truck could be equipped to support various teams. These variations in outfitting are often referred to by the military as loads, and the K-43-B had several, including K-43-C, K-43-D, and K-43-F.

The K-43 was most often used to outfit various crews, as shown below:

Hole digging crew
- (a) Earth Boring Equipment Crew – 5 men and 1 K-44-B
- (b) Hand Digging Crew – 8 men and 1 K-43 and 1¾-ton weapons carrier

Pole equipping, setting and guying crew
- 1 K-43 + 1 2-1/2 ton 6x6 cargo truck, 8 men

Wire stringing crew
- 1 K-43 + 1 K-36 trailer, 8 men
- or
- 1 2½-ton 6x6, cargo + 1 ¼-ton 4x4, 8 men

Wire tying crew
- 1 K-43 or 1 2½-ton 6x6, cargo, 8-men

As can be seen above, the number of men assigned to each crew exceeded the number that would fit in the cab of the

Above: *The winch drum and, to the left of it, part of the winch, are in view in the front end of the body of a Chevrolet K-43. The cover of the winch enclosure has been removed to permit a view of the winch and drum. To the right of the drum is the winch shaft extension, leading to the right side of the body, where the power reel CR would be installed when it was necessary to quickly lay down or, more frequently, reel in telephone cable. The winch was operated by a roller-chain drive from the transfer case.* General Motors LLC

Below: *On the lower half of the right side of the body of the K-43 were storage compartments. On the door of the forward compartment is a circular feature: this was the fitting at the right end of the extension of the winch shaft, upon which the power reel CR would be mounted when in operation. This fitting was a distinguishing feature for the K-43 as compared with the K-42, which lacked that fitting. On the front and the rear of the upper half of the body are holders for pike poles.* General Motors LLC

An early Chevrolet K-43 is viewed from the right side with the storage-compartment doors closed and the telescoping roof pulled to the front as far as possible. The extent to which the gin poles jutted out to the rear of the vehicle when in their stored position is apparent. General Motors LLC

1 ½-ton Chevrolet. From the outset, the Signal Corps had wanted trucks with five-man cabs. However, attempting to supply such a vehicle would have caused an unacceptable delay in production, therefore the Corps settled on the standard Chevrolet cab. In an effort to provide more seating capacity, the design of the truck included a padded seat and cushion, Chevrolet part number 607902, which was installed on top of the winch compartment. A similar seat, Chevrolet number 607907 was installed in the front of the YB truck beds as well.

K-44

The third of the trio of line equipment trucks built on the 1½-ton Chevrolet chassis was the K-44, the earth boring truck. Like the K-42 and K-43, the initial 81 trucks in this series had YP Chevrolet serial numbers. The Master Book symbol for the earth boring was YC. Produced in accordance with contract W-398-QM-8778, these G4112 series trucks were assigned Army registration numbers W-002852 through W-003114, and are covered in TM 10-1203 Change 1 June 15, 1941. The truck chassis for these vehicles were built in Norwood, Ohio, and shipped by rail to Cleveland for body installation by American Coach.

It is worthwhile to mention that both the K-44 and the K-43, when loaded, were very heavy, and for this reason the trucks were equipped with the same auxiliary spring arrangement as found on the 1½-ton 4x4 Chevrolet dump trucks. The earth boring machine alone weighed 2,500 pounds. It could bore holes 9, 12, 16 or 20 inches in diameter and 7'6" deep. Power for the boring equipment was supplied by the truck engine, driving through a special power take off. Also powered by the truck engine was a winch mounted in the front of the bed, which could be used with the mast of the boring machine to place, or remove, telephone poles. Like the winch on the K-43, the winch was neither designed nor well-suited for vehicle self-recovery.

The early trucks, known later as K-44-A, were equipped with the Type D earth borer, made by the Highway Trailer Company of Edgerton, Wisconsin. The earth borer bodies, the side panels of which were squared off at the upper rear corners, was based on the AT&T EBWP body design, and were produced by American Coach and Body. The

A power reel CR, also called a collapsible reel, is viewed close-up, positioned on the right end of the winch-shaft extension on a Chevrolet K-43. This reel was designed for rapidly paying out telephone wire or quickly taking up wire in neat, compact coils. A benefit of the power reel CR was that scrap telephone wire that had been taken up onto the reel could quickly be removed from the reel so that the reel could be used again immediately. General Motors LLC

The cab of the Chevrolet K-43 as viewed through the passenger's door was a very simple design. A single bench seat accommodated the driver and front passenger. The instrument cluster and controls were of the civilian type. Next to the driver's door was the winch clutch and brake-control lever. Rubber matting was laid on the floor, with lateral ridges for a nonslip surface. General Motors LLC

A Chevrolet K-43 is viewed from the front. The blackout lamps to the outboard sides of the service headlights are the early type without the rectangular aperture. The gin poles are seen protruding from the storage pocket in the left side of the body. General Motors LLC

Chevrolet K-43 registration number W-002909 was a 1941-model truck produced in Detroit under contract W-398-QM-8778. A caboscope rear-view device to assist the driver when performing pole-setting operations is on the cab roof. The gin poles are erected to form a tripod derrick. Highway-tread combat tires, size 7.50-20, are installed on the pierced-steel disk-type wheels. National Archives

American number for the body on these trucks was 5034-30 ACB. Eighty-one of these trucks were manufactured in 1941, with delivery beginning in April and continuing through September.

The K-44 was equipped with a Cab-O-Scope device, as described earlier in this chapter, which allowed the driver to monitor operations in the rear of the vehicle.

Like the K-42 and K-43, as originally produced there was a ten-inch gap between the rear of the cab and the front of the bed, and the fuel tank was mounted in this area. Not surprisingly, the 5 August 1941 Signal Corps Board report was very critical of this design.

Thus, subsequent production sported a completely redesigned bed, bearing little resemblance to the bed mounted on the YC trucks. The body design was modified in accordance with the recommendations of the 5 August 1941 Signal Corps Board. That recommendation was, "...practically identical with earth borer bodies now in service in the Army except that: (a) the rear top corners of the side panels have been rounded off to afford the truck driver better vision to the rear, (b) a cradle for supporting a single pole has been mounted on the top of the left side of the front panel of the body, and (c) the rear axle-to-end of body dimension has been fixed at 34 inches in order to obviate the possibility of excessive overhang."

The report went on to describe some of the differences between the K-44 body and the K-43 body (referred to as the LCW-102 body). The document reports, "The body of the Earth Borer and Pole Setter truck differs in design from that of the Telephone Construction Truck LCW-102 in two major respects, namely: the Earth Borer Body is neither equipped with side compartments above platform level nor a sliding roof. Two 36" steel side panels are provided in lieu of the side box panel construction characteristic of the LCW-102 Truck, and in addition, the body is approximately 9 inches shorter than the LCW-102 Truck body. Furthermore, the front panel of the Earth Borer Truck is equipped with two cradles, the one on the right for supporting the integral derrick while in the traveling position and the other on the left to support a single telephone pole should it be desirable to use the truck for hauling a single telephone pole."

The improved bodies, also produced by American Coach and Body Company, were given ACB number 5217-100. The vehicles based on the improved 1942 model chassis were assigned Chevrolet Master Book symbol and serial number prefix NR.

The 5 August 1941 report had pointed out the need to reinforce the rear of the frame on trucks equipped with earth boring equipment. Thus, when production of the K-44 resumed utilizing the NR chassis, the left and right side members of the frame were stiffened by reinforcement plates Chevrolet part numbers 3670563 (LH) and 3670564 (RH).

The vehicle, now referred to as K-44-B, was equipped with the improved Highway Trailer Company Type HD earth boring machine with power elevation of the rack shaft and boring case.

Deliveries of the revised trucks began in June 1942 under contract number W-398-QM-11508, also known as W-374-ORD-2744. This contract called for 154 of the vehicles, which were assigned registration numbers 007394 through 007518 and 007598 through 007986. Two of these trucks were delivered in June 1942, 30 in July, 25 in August, 34 in October and 63 in November.

Contract W-398-QM-11510 called for 522 more of the trucks, of which 312 were delivered in 1943 with the remaining 210 delivered the following year.

Contract W-374-ORD-2839 included two groups of K-44 trucks. The first group, consisting of 615 vehicles, were delivered 77 in 1944, and 538 in 1945. The second group, which were also delivered in 1944, amounted to 287 vehicles.

Just as the K-43 had various loads corresponding to crew duties, including K-43-C, K-43-D, and K-43-F, so too did the K-44. In addition to the previously discussed K-44-A and K-44-B variations, the vehicle also existed in the K-44-C, K-44-D, K-44-E and K-44-F configurations.

The same K-43, W-002909, demonstrates its ability to hoist a telephone pole using the tripod derrick. In the background are several rows of telephone poles that have been erected. It was necessary to slide the roof of the body forward when using the winch cable. National Archives

Chevrolet K-43 registration number W-002909 is seen from the rear with the derrick erected. The forward, or center, gin pole of the derrick was attached to a fitting on the floor of the body, while the two rear poles were fastened to fittings on the rear of the body. A tube with a sheave at the center is mounted on brackets on the upper rear of the body. The winch is visible in the forward part of the body. The power reel CR is stored on the left side of the interior of the body. National Archives

This K-43 cab has been equipped with controls for a brake system for a towed trailer. The equipment, made by the Warner Electric Brake Manufacturing Company, includes a control knob to the left of the instrument cluster and a control lever mounted on the left side of the steering column. National Archives

Lend-Lease documents show that 19 K-44 trucks were supplied to the British Empire and 16 to the Free French. The British trucks were assigned registration numbers L6229329 to L6229348.

It is notable that in the Statistical Work Sheets prepared by the Office Chief of Ordnance – Detroit, Engineering-Manufacturing Division, Requirements and Progress Branch of August 1945, the K-43 and K-44 were the only 1½-ton 4x4 Chevrolet types scheduled for production beyond July 1945. Planning at that time would have included the following:

Total planned 1945 production K-44 693
Total planned 1946 production K-44 177
Total planned 1945 production K-43 657
Total planned 1946 production K-43 607

Production of the K-44 was scheduled to continue through March 1946, with the K-43 remaining in production through September. These trucks were not only forecast, but contracts and production orders had been issued.

Of course, there would be no 1946 production of either type, and the Office Chief of Ordnance – Detroit document Summary Report of Acceptances Tank-Automotive Material records that only 538 K-44 trucks were built in 1945, and 499 K-43 trucks were built in 1945.

What makes this all the more remarkable is the harsh criticism that was leveled at these vehicles. The Army Service Forces, Signal Corps Engineering Laboratories, Coles Signal Laboratory, Red Bank, New Jersey – Report of Wire Equipment Committee, 11 October 1945 – "In pole line construction, the methods and equipment used for rural telephone lines in this country, such as Class 9 poles and Truck K-44 pole setting equipment, are still in use by

All of the right-side storage compartment doors are open on Chevrolet K-43 W-002909, and the derrick is assembled on the rear end. The beverage container is visible in the upper rear compartment. Sliding trays for parts are in the two compartments above the rear wheel well. National Archives

the Army. Trucks K-43 and K-44 are listed as standard items on Tables of Equipment. However, experience in the European Theater indicated that these trucks are not satisfactory, either as a motor vehicle or as a piece of Signal Corps equipment. At the close of the war in that theater, one Field Army had discarded both vehicles referred to in favor of the standard 2-1/2 ton, 6 x 6."

"VI. Required Research and Development……

c. Pole Erection. Truck K-44 pole hole digging equipment is unsatisfactory in all respects. A completely new and much more rugged vehicular pole hole digging equipment, designed around a standard Ordnance or Corps of Engineer's chassis is required."

The Corps of Engineers, who also used the K-44 under their own designation of Earth Auger Truck M1, in 1944 had begun developing CCKW-based 6x6 replacements for both the K-43 and K-44. The Engineers had begun developing these for use in power line construction and maintenance, however the Chief Signal Officer directed that the Signal Corps Board cooperate with the Engineers in developing these vehicles, so as to make the resultant trucks jointly useable. By June 1945 conversion kits to mount the necessary specialized bodies and equipment for both Earth Boring and Line Maintenance (V-18 and V-17 respectively) on CCKWs had been perfected and were recommended for Standardization. This paved the way for the Chevrolet-based vehicles to be reclassified first as Substitute Standard, and later as Obsolete.

Above: *Chevrolet K-43 Army registration number 007213 was a 1942-model vehicle, built at Cleveland, Ohio, under contract W-398-QM-11508 (later, W-374-ORD-2744). Several noticeable changes are visible on this vehicle compared with the early-production K-43s include four rivets instead of six to secure the front bumper to the chassis frame; a redesigned bracket for the hand-starting crank on the top center of the bumper; redesigned windshield washers; and marble reflectors. Also, the gap between the cab and the body was greatly reduced, as the body had been lengthened, and the fuel tank now was inside the left front of the body; the fuel filler protruded through the front of the body.* General Motors LLC

Below: *The roof over the body of Chevrolet K-43 007213 is shown in the fully extended position, covering the entire body. There were three telescoping roof sections. The front and the rear sections appear to have been rigid, with lateral ribs, while the center section was of canvas or fabric without lateral ribs.* General Motors LLC

Above: *A left-rear view of K-43 007213 shows the gin poles stored in the pocket in the left side of the body. Marble reflectors are on the side and rear of the rear corners of the body. The metal tube with the sheave in the center, for winch-cable operations, is stored on the brackets for the rear gin poles.* General Motors LLC

Below: *Partially surrounding the tail lights under the rear of the body of the K-43 were U-shaped crew steps, with a bent piece of diamond-tread steel attached to each one. A large eyebolt was mounted on the center of the rear of the floor of the body, partially hidden in this photo by the tube with the sheave.* General Motors LLC

Above: *Chevrolet K-43 Army registration number 007213 is observed from the right side. Another design change that differentiates this vehicle from earlier K-43s is the new position of the battery, below the rear of the cab and the front of the body, with cutouts made in the running board, the apron, and the body to provide clearance for the battery and fuel filter.* General Motors LLC

Below: *The front of Chevrolet K-43 007213 is displayed. The blackout marker lamps now had rectangular apertures on their front faces. The bracket between the tow hooks on top of the bumper had been redesigned; the early type had the flange at the bottom pointing to the rear; this new type had the flange facing the front, and the front face of the starting crank support was taller.* General Motors LLC

Above: *Chevrolet K-43 registration number 0099797 is seen on 9 May 1945 before the start of testing of the vehicle at the Ordnance Operation, Studebaker Proving Ground. This vehicle was a model G-7173, built under contract number W-374-ORD-2839 and accepted by the Army during 1945. Stenciled on the cab door is "E-2-96," the vehicle's Studebaker Proving Ground number.* TACOM LCMC History Office

Below: *The U.S. Army Signal Corps ordered large numbers of the Chevrolet Pole-setter Truck with Earth Borer K-44. Based on a Chevrolet 1 1/2-ton 4x4 chassis, this vehicle was equipped with a winch in the specially designed truck body and a rear-mounted combination earth borer and derrick, with which a crew could bore holes in the ground and set poles. In this photograph of a 1941-model K-44, registration number W-002853, the derrick is lowered to the travel position, its top being visible above the cab.* General Motors LLC

Above: *As seen in this left-side view of Chevrolet K-44 W-002853, the fuel tank was located between the cab and the body. Two storage compartments were toward the bottom of the side of the body, with doors with a piano hinge on the bottom and a single T-shaped latch handle. To the rear of the body is the earth borer, with an auger attached. On the early K-44s, the top rear corner of the body was squared. Later, the body would receive a rounded top rear corner, and this would be referred to as the "new" body, to differentiate it from the "old" body.* General Motors LLC

Below: *The borer and auger, as seen in the lowered position in the rear of K-44 W-02853, were mounted near the centerline of the body. The control levers for the borer are visible just above the "U.S.A." stenciled on the rear of the body. One lever controlled the up and down feed of the auger, and the other controlled the drive to the auger. The spare tire is stored on the right side of the interior of the body. The horizontal tubular structure to the front of the borer is the main supporting case tube, which supported the borer case and housed the main-drive and feed shafts.* General Motors LLC

Above: *As seen in a rear view of the same K-44, on each side below the body of the K-44 was a pole-derrick support, similar to a swiveling jack. These are seen in the travel position. When boring in heavy soil such as clay, the supports were deployed to take the load off the suspension springs. On the left wall of the body is stored a power reel CR, which would be attached to the winch shaft on the right exterior of the body in order to quickly play out or reel in telephone cable.* General Motors LLC

Below: *In an elevated view of the rear end of K-44 W-002853, in the foreground to the left of the auger is the intermediate case, with the boring case to its right. Atop the intermediate case is the leveling worm that effected the fore-and-aft adjustment of the derrick; a ratchet wrench is attached to that worm. On the opposite (or top) end of the intermediate case is the leveling worm that operated the left-and-right adjustment of the derrick. Together, the two leveling worms were adjusted to make the derrick plumb before boring a hole for a telephone pole. In the front of the truck body is the 18-inch winch drum.* General Motors LLC

Chevrolet K-44 W-002853 is observed from the rear with the derrick and the borer erected and the derrick supports deployed in their lowered positions. The derrick was raised to the operating position by one crewman turning the fore-aft leveling worm and another crewman who stood in the truck body pushing up on the derrick. The earth borer was operated by the truck engine via the transmission and the power takeoff. Military History Institute

Above: *In a view of the right side of Chevrolet K-44 W-002853, the round feature toward the front end the truck body is the shaft from the winch drum, onto which the power reel CR was mounted when in operation. To the front of the body is the fuel tank, on top of which is the filler and cap, tilted toward the right side of the vehicle.* General Motors LLC

Below: *Partway through Chevrolet K-44 production, a feature called the Cab-O-Scope was added to the left side of the cab roof. This consisted of a large rear-view mirror in a hood-shaped housing. The Cab-O-Scope enabled the driver to more easily see the earth-borer operator's hand signals when he called for more engine power or transmission speed. A Cab-O-Scope is present on the cab roof of this K-44, registration number W-003041.* National Archives

The cab of an early K-44 is viewed through the passenger's door. A civilian-type instrument cluster was provided. On the glove box door are the serial number plate and a data plate listing maximum permissible road speeds. On the left side of the floor, next to the driver's door, is the winch lever. The nearest lever on the floor is the power takeoff control lever. When boring earth, the borer operator signaled the desired speed to the truck driver, who placed the power takeoff lever into the earth-boring position and then regulated the transmission speed and the engine speed per the borer operator's instructions. General Motors LLC

When viewed from the front, early K-44s are distinguishable from later specimens by the lack of a Cab-O-Scope on the roof of the cab over the driver's station. The derrick is in the travel position in this photograph. When viewed from the front, early-production K-44s are distinguishable from later specimens by the lack of a Cab-O-Scope on the roof of the cab over the driver's station, the early louvered blsckout marker lights and center canted tow hooks. The derrick is in the travel position in this photograph General Motors LLC

This closeup photo of the right side of a K-44 shows the 35-gallon fuel tank, bolted to the front of the truck body. A wire mesh grille is fitted over the opening in the front of the body, to protect the rear window of the cab. A good view is available of the terminus of the winch shaft in a round recess on the side of the body; this is where the power reel CR would be installed when in operation. National Archives

Above: *The earth borer on K-44 registration number W-003041 is set to dig a hole in the ground at approximately 45 degrees from horizontal. A retouched version of this photograph was used to illustrate the Truck K-44-B with Earth Borer Equipment HD in the Army technical manual TM 11-364,* Truck K-44-B and Earth Borer Equipment HD, *dated September 1942. The K-44-B was the version of the truck with the Type HD earth-borer equipment, while the K-44-A had the Type D earth borer.* National Archives

Below: *Chevrolet K-44 W-003041 is shown with the derrick and borer deployed for drilling a hole in the ground. Visible on the roof is the Cab-O-Scope, and atop the front of the truck body is a holder for the derrick when lowered to its travel position. A sheave is at the top of the derrick, and another sheave is on the rear side of the base of the derrick; these were for handling winch cables during the hauling of telephone poles into position for insertion into the ground.* National Archives

Above: *This vehicle, registration number 007394, represented a late-type K-44, which had a redesigned truck body (often referred to as the "new" body), the sides of which were rounded at the rear. The gap between the cab and the front of the body was narrowed to 1.5 to 2 inches; the fuel tank was removed from that gap; and a new, 35-gallon fuel tank was installed in the left front of the body, with the filler neck protruding from the left side of the front of the body. This 1942 model-year vehicle was built in Cleveland under contracts W-398-QM-115408 and, later, W-374-ORD-2744.* General Motors LLC

Below: *The rear of the body was rounded off to offer the driver better vision to the rear. The dimension from the rear axle to the rear of the body was fixed at 34 inches; it was determined that a longer dimension would have placed overload strain from the heavy borer assembly onto the overhang at the rear of the body.* General Motors LLC

Above: *The earth borer is protruding from the rear of the late-production body of K-44 registration number 007394. In addition to the cradle for holding the derrick in the travel position, another cradle now was mounted on the left top of the front of the body, to support a single telephone pole. Marble reflectors are mounted on the sides and rear of the body.* General Motors LLC

Below: *In a left-rear view of K-44 registration number 007394, the earth borer and derrick are in the travel position, the derrick resting on its cradle. The cradle for a telephone pole is visible to the left of the derrick cradle. The power reel in its stored position and a large auger are visible in the left side of the body. The spare tire and, to its rear, a beverage container are stored on the right side of the body.* General Motors LLC

Above: *In another rear view of K-44 007394, the newly repositioned fuel tank is visible in the front right corner of the body, partially obscured by the stored power reel CR. The intermediate case of the earth borer—the case to the left of the borer case to which the auger is attached – now has a drain plug on its rear face, not seen in photos of the earth borers on early-type K-44s. General Motors LLC*

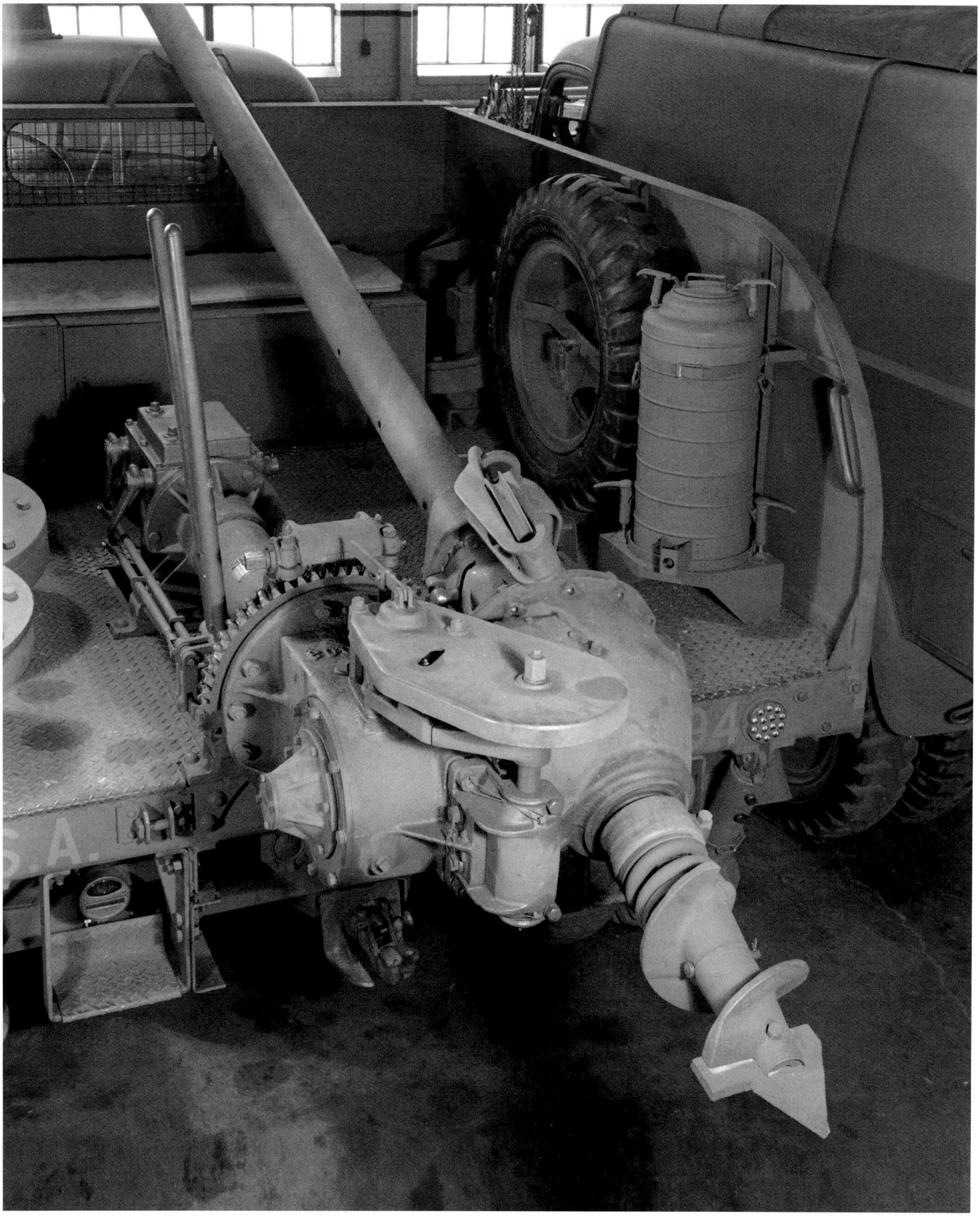

A Type HD Earth Borer on a late-production K-44 is viewed close-up from the rear. In addition to the drain plug on the rear face of the intermediate case, a new feature was the metal cover mounted over the top of the intermediate case. Protruding through the top of the case is the fitting for adjusting the fore-and-aft worm mechanism by means of a detachable wrench. A small-diameter augur bit is installed on the borer. On the floor to the front of the borer are the case and linkages for the main drive gear and the drive and feed clutches. General Motors LLC

In a view of an earth borer in a K-44 from the front, the nomenclature plate on the main supporting case tube in the foreground identifies the unit as a Type HD Earth Borer, manufactured in late March 1942. Mounted to the floor between the supporting case tube and the derrick is a sheave used with the winch cable in conjunction with the other sheaves on the top and the bottom of the derrick. The sheave at the bottom of the derrick is visible here, connected to a U-bolt. General Motors LLC

Above: *Chevrolet K-44 registration number 007394 is viewed from the right side. For these later K-44s, the battery was repositioned to below the junction of the cab and the body, with cutouts in the cab and body to give clearance to the battery. Whereas the early-production K-44 bodies had single-panel doors with piano hinges on the bottoms for the forward storage compartments, late-model bodies had two-panel doors with vertical piano hinges. A marble reflector is on the door of the rear storage compartment.* General Motors LLC

Below: *As seen in this frontal view of K-44 007394, the fuel filler (partially obscured here by the rear-view mirror) extended well to the outboard of the side of the truck body. The grille was of accordion- or zig-zag-type bent metal rod construction.* General Motors LLC

Above: *Chevrolet K-44 Army registration number 0080961 and manufacturer's serial number 9NS31-10458 was photographed for the Ordnance Operation, Engineering Standards Vehicle Laboratory, Detroit, on 5 September 1944. In the serial number, "9" indicates that the truck was constructed at the Norwood, Ohio, plant; "NS" refers to telephone truck with winch; "31" translates to July 1944 production; and "10458" signifies that this was the 9,458th vehicle of the production run lasting from August 1943 to July 1944. (At the start of August each year, the final element of the manufacturer's serial number reverted to 1001.) Improvements visible in this photo include a pioneer tool rack on the front of the body and five-gallon water can and holder on the running board.* TACOM LCMC History Office

Below: *K-44 registration number 0080961 is viewed from the right side in a 5 September 1944 photograph. Although barely visible, "CHEVROLET" is embossed on the side panel below the hood, to the front of the louvers. Signal Corps specifications preferred the use of aluminum paint on the interior of vehicular storage compartments, such as those on the K-44, so that parts in the compartments were more readily visible. However, in light of the wartime shortage of aluminum paint, light gray paint was used on the storage compartments of these trucks.* TACOM LCMC History Office

Above: *In this overhead view of K-44 registration number 0080961, a close inspection reveals that the two control levers for the earth borer were hollow pipes. Two large augers of different sizes are stored on the left rear side of the body, while two smaller augers are stored on the forward end of the right side of the body, flanking the right extension of the winch shaft.* TACOM LCMC History Office

Below: *Chevrolet K-44 registration number 0080961 is observed from the rear. No auger is installed on the borer in this photo. By now, a beverage container was a regular part of the on-vehicle equipment, secured to a holder in the right rear of the body. Note the grab handle on rear edge of each side of the rear of the body.* TACOM LCMC History Office

Above: *In a left-rear view of K-44 0080961 dated 5 September 1944, the cogged worm wheel for leveling the borer from side to side is prominent on the front end of the intermediate case of the earth borer. The derrick supports are collapsed, tilted forward, and secured with chains for travel.* TACOM LCMC History Office

Below: *As seen on K-44 0080961, the left forward storage compartment of the body, like the forward compartment on the right side, was equipped with two side-by-side doors, each secured to the body with a vertical piano hinge. One T-shaped latch handle served both doors.* TACOM LCMC History Office

A frontal view of K-44 0080961 provides details of the grille, the brush guards, the grille on the rear window, the fuel filler on the left front of the body, and the pioneer tool rack on the right front of the body. TACOM LCMC History Office

Compared to the photo on a preceding page of the interior of the cab of an early-production K-44, the cab of this late-production K-44, Army registration number 0080961, has a military-type instrument cluster, lights, and switches; the maximum-speed warning data plate is no longer on the left side of the glovebox, and a universal rifle bracket has been installed on the right side of the cab. TACOM LCMC History Office

Chapter 5

Truck Tractors

Perhaps one of the most misunderstood of the 1½-ton 4x4 Chevrolets is the truck tractor. Much of the confusion, at least when viewing photos, is the result of not understanding the Government procurement methodology early in the war.

While the fifth wheel coupling is to many people the identifying characteristic of a truck tractor, for most of the war this key item was not a part of the truck tractor at all, but rather, a component of the trailer. In this regard, the Army was following civilian practice. Fifth wheel couplers are often installed after the truck-tractor leaves the factory. In the Chief of Ordnance – Detroit document "Engineering of Transport Vehicles, 1942-1945," it was recorded "Prior to 1 January 1944, it was general practice to procure truck-tractors less lower fifth wheel and trailer brake and electrical jumper connections. These items were procured with each semi-trailer and installed on the truck-tractors when the two vehicles arrived at the same delivery point. This practice resulted in

Chevrolet produced a truck-tractor version of the G-506 1 1/2-ton 4x4, successively assigning to it Chevrolet model numbers G-4112, G-4165 (late 1941), and G-7113 (1942 and later). The chassis had a 145-inch wheelbase and was equipped with a fifth wheel for coupling a semi-trailer. Before 1 January 1944, Chevrolet shipped the truck-tractor's fifth wheel and electrical connections as separate components with the semi-trailers, and these items were installed on the truck-tractors when those vehicles were paired with semi-trailers at the delivery point. Shown here at Holabird Quartermaster Depot, Baltimore, Maryland, around October 1941 is a 1941-model-year G4112 truck-tractor, registration number W-328765, constructed under contract W-398-QM-8686. The fifth wheel has not been installed. National Archives

1941 Chevrolet 1 1/2-ton 4x4 truck-tractor W-328765 is viewed from the right side. Features worthy of notice are the mud guards to the fronts of the rear tires; the filler neck for the fuel tank, to the rear of the cab; and the toolbox with a padlock along the chassis to the rear of the cab. The tires are Goodrich Silvertown, marked "TRUCK-BUS / SUPER TRACTION." General Motors LLC

a surplus of lower fifth wheels and jumper connections, as it is the usual practice to procure more semi-trailers than truck-tractors. Further, lower fifth wheels were being installed by many different depots, base shops and ports of embarkation, resulting in variation in mounting and positioning on truck-tractor chassis, and unnecessary work and loss of time at these locations." While production of the 1½-ton Chevrolet stopped prior to factory installation of the fifth-wheel becoming standard, the above passage does provide some insight as to the circumstances under which the vehicle was produced.

The 1½-ton Chevrolet tractor was intended to tow the two-wheel semi-trailer, 3-ton van or 3½-ton stake and platform. Both types were 16-feet long, and had a gross weight of six tons.

The 16 September 1940 proposal to produce trucks, as well as the 24 October 1940 contract for the production of trucks, included a truck-tractor variant of the Chevrolet. That contract, W-398-QM-8686, called for eight of the tractors to be built. These trucks, while using the serial prefix of YP, were assigned the Master Book Symbol YK and US Army registration numbers of 328761 through 328768 inclusive.

Contract W-398-QM-9402 included 89 more tractors, all of which were delivered in 1941. Beginning in April 1941 through September 1941, Chevrolet used serial prefix ZP on these trucks, while also assigning Master Book Symbol ZK. These trucks bore US Army registration numbers 337946 to 338034 inclusive.

Next came two concurrent contracts, DA-W-398-QM-8 issued for 24 more examples for Defense Aid, and W-398-QM-10328 for 48 tractors for US Army. The latter group of tractors was assigned Army registration numbers 347708 through 347740 inclusive, 353229 through 353242 inclusive, and 353243. All of these, as well as the Defense Aid trucks were delivered in 1941.

Truck-tractors contracted for after August 1941 used both serial number prefixes and Master Book Symbol NK. Tarrytown built 39 more truck tractors for Defense Aid on contract DA-W-398-QM-252; these bore the NK designator.

The same truck-tractor, registration number W-328765, is viewed from the front. The blackout marker lamps on the fenders of this vehicle, on the outboard sides of the service headlights, are the early, "louvered," type. General Motors LLC

The cab of a Chevrolet 1 1/2-ton 4x4 truck-tractor is viewed through the right door. The instrument cluster is the civilian type. These truck tractors were equipped with electric trailer brake controls, clamped to the steering column, and this feature is visible just below the steering wheel. Installed in the dash at the far left side is a rheostat to adjust the sensitivity of the electric braking. General Motors LLC

A truck-tractor, likely registration number W-328765, is seen from the rear, showing the dual rear wheels and early louvered "cat's-eye" type tail lights. Also of interest are the angle-iron holders for the metal mud flaps, called quarter-fenders, visible above the tires. In service, the truck-tractors were equipped with stationary fifth wheels for coupling a semi-trailer, but early in the war the fifth wheels were supplied with the trailer, not the tractor. General Motors LLC

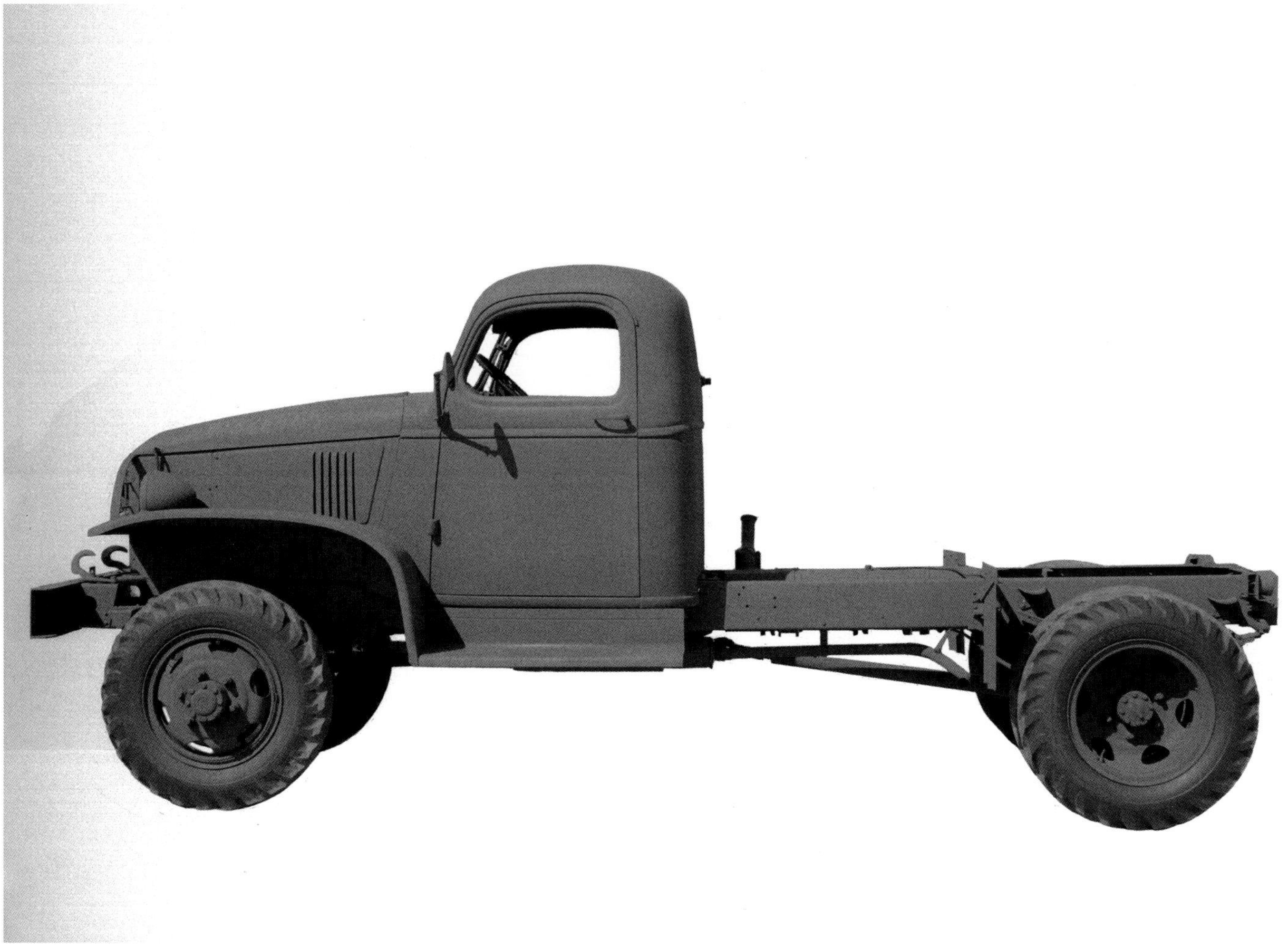

The left side of 1941 Chevrolet 1 1/2-ton 4x4 truck-tractor W-328765 is displayed. Although this feature is very difficult to discern, "CHEVROLET" is stamped on the side panel of the hood, near the upper front of the six ventilation louvers. The Army ordered the discontinuation of the practice of stamping manufacturers' names on its vehicles inin late 1942. General Motors LLC

During 1942, a further 44 tractors were delivered on contract W-398-QM-10939. But it was the 4 January 1942 order; contract W-398-QM-11510, that procured the bulk of the truck tractors, with 2604 examples ultimately on the order.

All the tractor trucks featured auxiliary springs, also known as 'overload' or 'helper' springs. Such springs were also used on the long wheelbase trucks, telephone and earth boring trucks, and dump trucks. Regardless of application, the auxiliary spring packs were the same. However, the bracket, which the springs contact when under load on the ZK and NK trucks, differed from that used on all the others.

Deliveries of the truck tractor ended with 56 examples in March 1944. The report "Recapitulation of Facility Expansions for Tracked & Wheeled Vehicles from 1940 to July 15, 1945," prepared by the Special Planning Branch, Office of Chief of Ordnance – Detroit lists that a total of 2,935 truck tractors were delivered to the US Army.

Lend-Lease documents show that the sole foreign recipient of the 1½-ton 4x4 tractor under those terms was French Forces, which received 49 units. In a January 1945 report on the Army Supply Program, it was noted that average cost for each of these truck tractors was $1,533.60.

Above: *Chevrolet 1 1/2-ton 4x4 Truck-tractor registration number 337947, also with the fifth wheel not yet installed, is viewed from the right side. This vehicle was built as a Chevrolet model G-4165 during the 1941 model year under contract W-398-QM-9402. The spare tire on the truck-tractors was relocated to the rear of the cab and was stored vertically on a carrier. This tractor has the later frame-mounted battery with cutout in the body.* General Motors LLC

Below: *A different Chevrolet 1 1/2-ton 4x4 Truck-tractor is viewed from the left rear. This is a later tractor with the battery mounted on the right side of the frame, making it necessary to move the tool box to the left side frame rail. This vehicle also has a pintle hook installed in the rear cross-member. The fuel filler is visible to the rear of the spare tire.* General Motors LLC

Above: *In a photo taken for Holabird Quartermaster Depot in Baltimore and dated 24 January 1942, Chevrolet 1 1/2-ton 4x4 Truck-Tractor registration number W-338013 is coupled to a Winter Weiss six-ton (gross) semi-trailer. This truck-tractor has the early, symmetrical continuous bent rod brush guard and "CHEVROLET" stamped on the side panel of the hood.* TACOM LCMC History Office

Below: *A Ward Semi-trailer Ambulance is coupled to a Chevrolet 1 1/2-ton 4x4 Truck-tractor. The registration number of the truck-tractor is not legible, but a registration number, 0490148, is stenciled on the right rear of the ambulance below "U.S.A." The insignia of the U.S. Army Medical Department is on the side of the ambulance near the front.* TACOM LCMC History Office

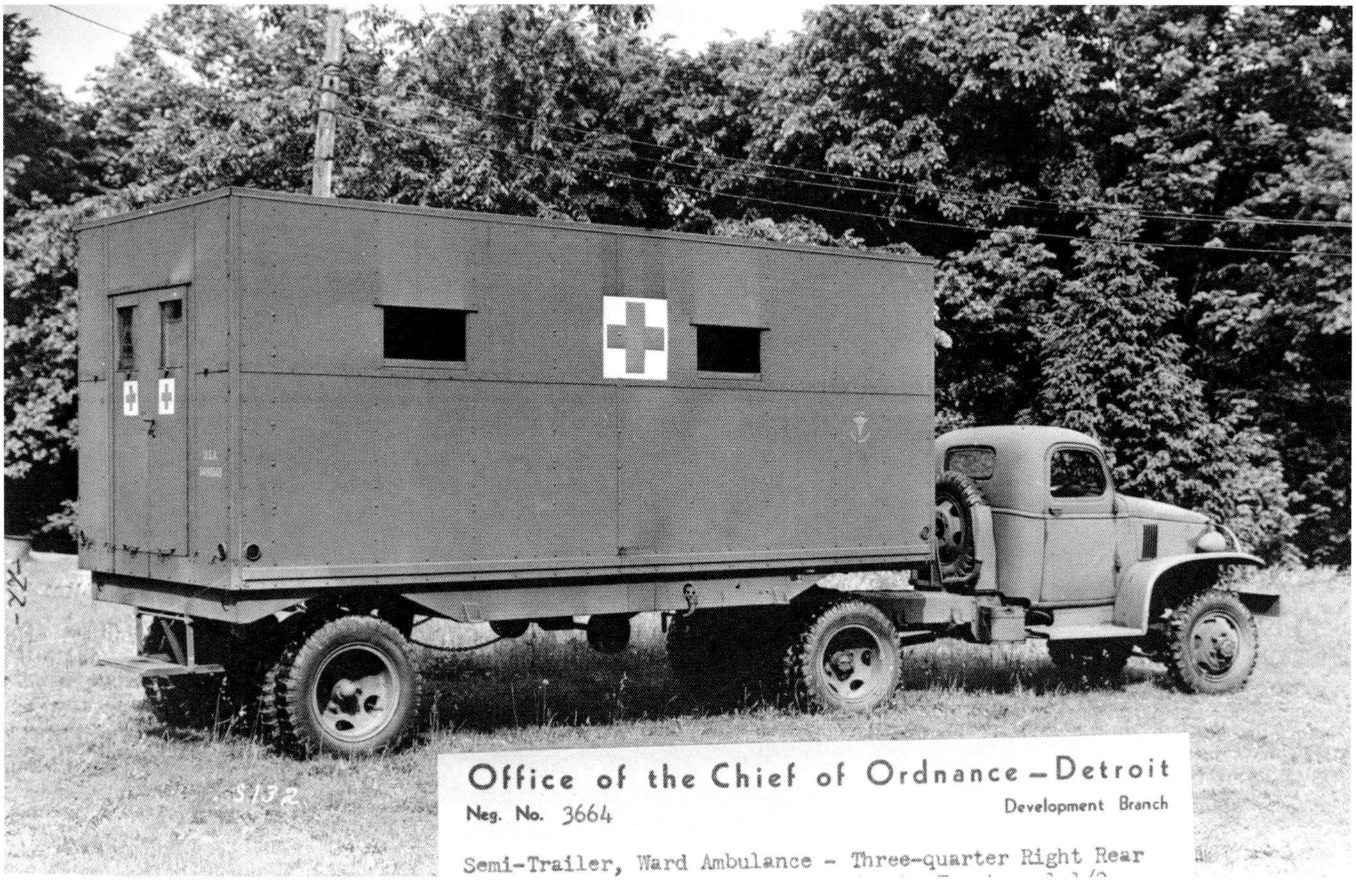

Above: *This Detroit-built G-506 Chevrolet, registration number W-338013 or W-338015, was built and delivered as a truck-tractor under contract W-398-QM-9402, with acceptance in 1941. The fifth wheel is installed, and a spare-tire carrier is visible behind the right side of the cab.* National Archives

Below: *Vintage adverts.* General Motors LLC

Above and Below: *Vintage adverts.* General Motors LLC

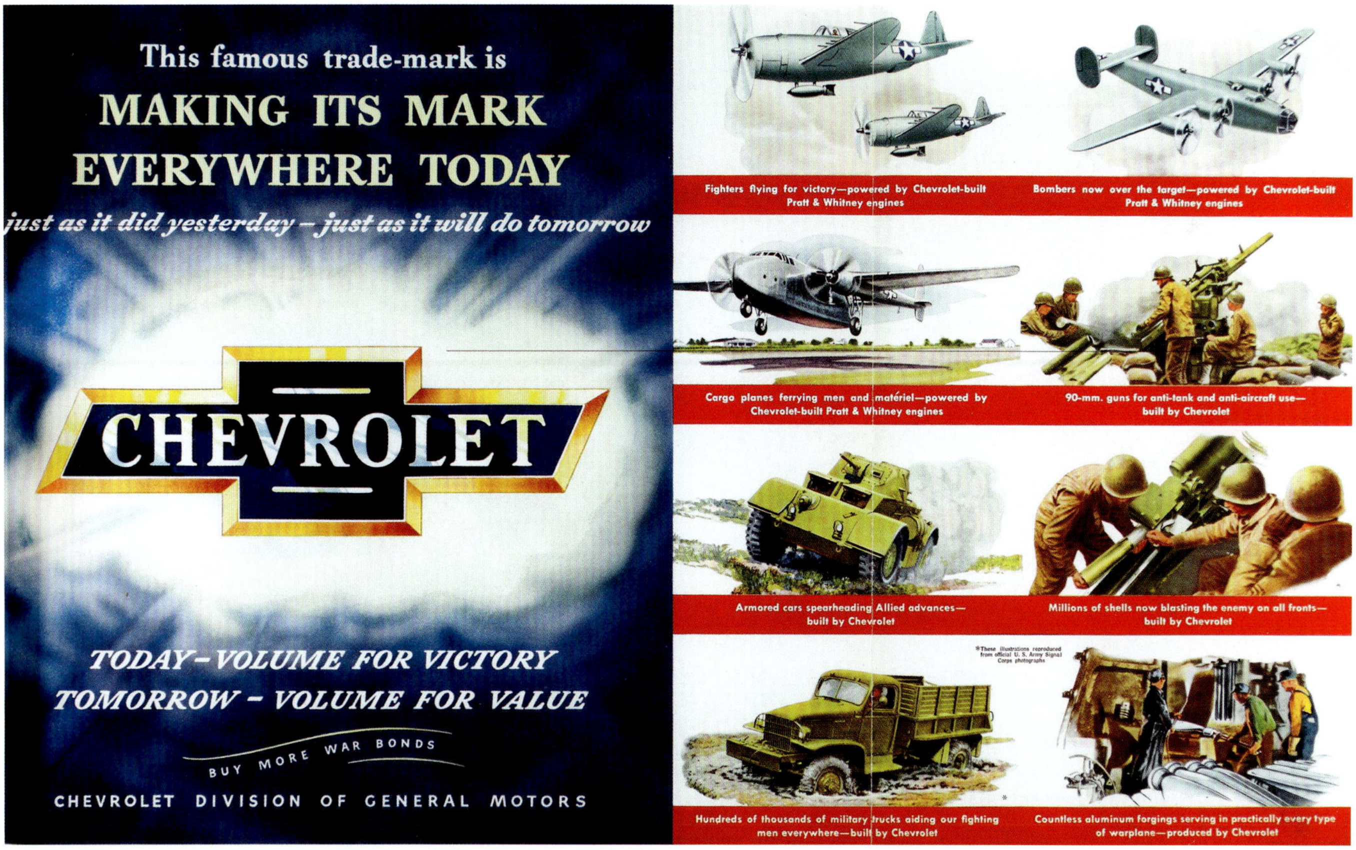

Chapter 6

Cab Over Engine

The Signal Corps, often in the need to transport unusually long items such as antenna components, had a tendency to prefer cab over engine (COE) trucks, which allowed slightly larger cargo beds to be used. The Chevrolet 4x4 was no exception to this, and the 17 August, 1940 specification for the truck included "Trucks, 1 ½-ton, (light chassis), motor, gasoline, C.O.E., with stake and platform bodies." These trucks were seen as a less expensive alternative to the Autocar 4-ton COE trucks already slated. The 1½-ton Chevrolets were given the Signal Corps designation K-33, and were procured specifically for transporting antenna components.

Contract W-398-QM-8686 issued 24 October 1940 included 96 (some documents say 94, however both the "Summary Report of Acceptances of Tank-Automotive Material" as well as the report "Recapitulation of Facility Expansions for Tracked & Wheeled Vehicles from 1940 to July 15, 1945," prepared by the Special Planning Branch, Office of Chief of Ordnance – Detroit, list 96) of these trucks, all of which were delivered in 1941. The trucks used serial prefix YX, and Master Book Symbol YN. Army registration numbers assigned were W-328667 though W-328760. The first six of these trucks were delivered in January 1941, with 41 more in February; 36 in March and the final 13 in April.

No COE platform and stake body trucks were built on the late 1941 (Z-series) production, but the type did reappear in the 1942-model production trucks. First was contract W-398-QM-10708 for 60 of the trucks. Built in Baltimore, they were assigned registration

A cab-over-engine (COE) model was produced in the G-506 Chevrolet line: originally designated the G-4103, it known as the G-7123 from 1942 on. The Signal Corps used the designation K-33 for these trucks. Features of this model included a snub-nosed, enclosed cab; a 175-inch wheelbase; and an identical drive train to the other trucks in the G-506 line. A separate brush guard was provided to protect the headlights and grille. National Archives

Above: *The G-7123 cab-over trucks had a 16-foot-long cargo body, the back end of which jutted far to the rear of the rear wheels. The cargo body was equipped with stake sides all around. The headlights were installed in housings that fit the curvature of the front fenders.* General Motors LLC

Below: *A G-7123 is observed from the left rear. The side stakes have a gate built into the middle. Reflectors are suspended below the rear corners of the body. Below the rear of the body are two bumperettes, with a tow pintle between them.* General Motors LLC

Above: Below: *Access to the cab was facilitated on both sides of the G-7123 by a small running board, above which was an even smaller step, attached to the rear of the front fender. A tow hook is visible below the front bumper.* General Motors LLC

Below: *In a photograph of the components involved in transporting the SCR-270-B Early-Warning Radar equipment, to the left is a Chevrolet K-33 Stake Body Truck. This was the Signal Corps designation for the Chevrolet 1 1/2-ton 4x4 COE Truck. To the front of this truck are an Autocar K-32 Prime Mover pulling a K-22-B Antenna Trailer; an Autocar K-31 Power Truck; and an Autocar K-30 Operating Truck.* US Army Communications-Electronics Command

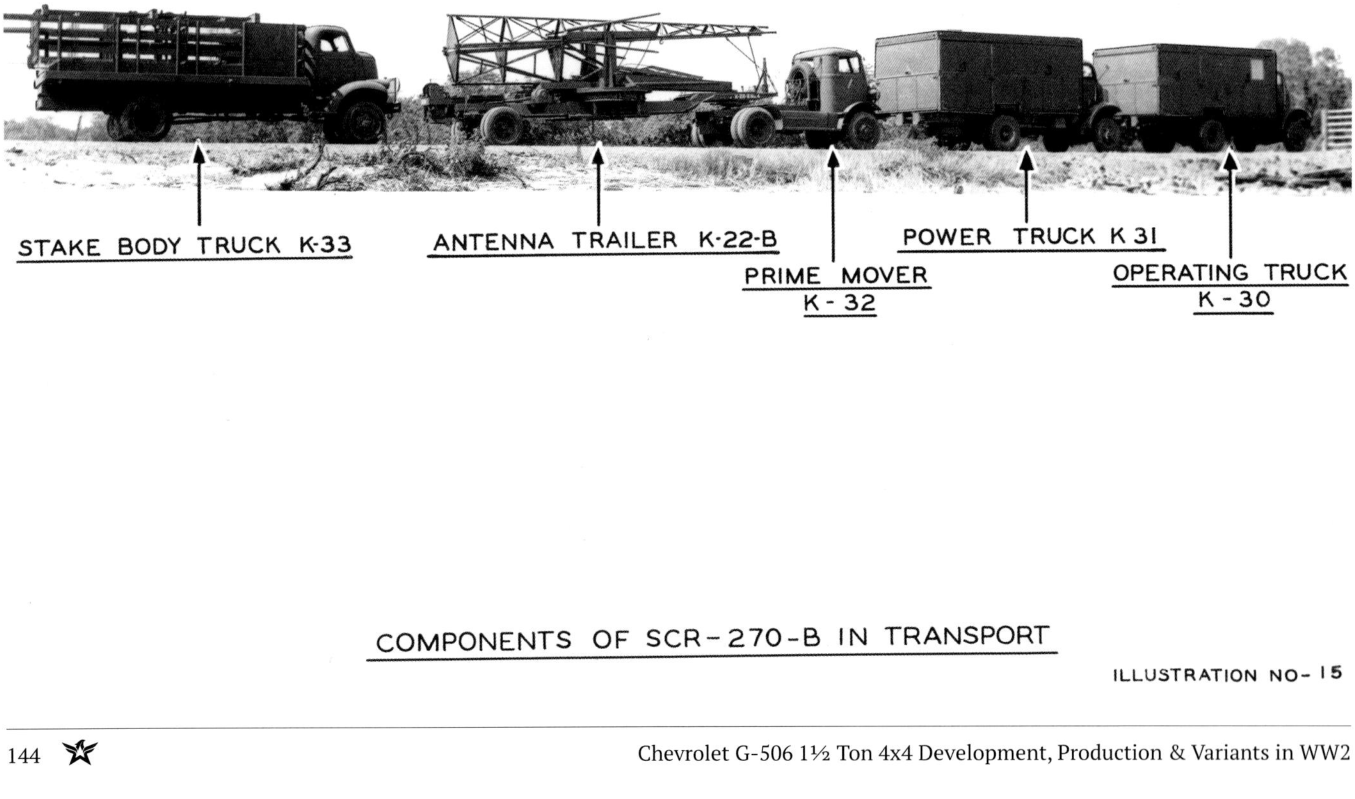

The final variant of the cab over engine G-506 was the K-54, which was procured by the Signal Corps to transport SCR-270 radar set components. The K-54 cargo body was slightly longer than the K-33 body, and it had very tall bows, resulting in an overall height of 117-inches.

numbers W-343779 through 343838 and were delivered in February (35), March (15) and April (10) of 1942. These and all subsequent cab over engine stake and platform body trucks used serial prefix and Master Book Symbol "NN."

Contract W-398-QM-13383 included 357 COE trucks that were built in Norwood. These trucks were assigned registration numbers W-3114446 through W-3114802.

The final contract to include the COE stake and platform body trucks was W-398-QM-11510, which called for 85 of the vehicles. These trucks were assigned Army registration numbers 3219414 through 3219498.

Deliveries of the vehicles on contracts W-398-QM-13383 and W-398-QM-11510 began in October 1942 and continued through May 1943, with the exception of December 1942, when none of the type were delivered.

While the 193-inch long cargo body of these trucks was longer (and 19 ½-inches wider) than the 180-inch long bed of their contemporary, the long wheelbase conventional cab G-506, the decision was made to create an even larger bed. A standard component of the SCR-270 long distance early warning radar set, even the K-33 bed was not big enough to completely enclose the antenna. Thus, the K-54 was created, which was 4½-inches longer as well as taller, in order to accommodate the oversize cargo. Some evidence suggests that the 150 trucks completed on contract W-398-QM-13383 in 1943 were finished in this configuration.

Chapter 7

Cab and Chassis

In addition to the various types of vehicles that left the Chevrolet plants as complete trucks, or which had their chassis assembled for a specific purpose, such as the telephone trucks, the Army also purchased plain cab and chassis. Upon these chassis could be installed various types of bodies.

While most of the Chevrolet records relating to wartime production have been lost, leaving us with few insights as to why and how that division did things, the records of sister GM division GMC are more complete. Those records indicate that in May 1943 the Detroit Ordnance District agreed to accept uniform chassis types, rather than chassis specific for a particular type of future bed installation. That is, previously certain details such as spare tire mounting, taillight and trailer electrical receptacle, etc. would vary dependent upon proposed future use. Such specialization greatly complicated production, as well as locking in a particular chassis for particular use months before it actually began assembly. Universal chassis provided the government with more flexibility, as well as streamlining production.

Based on the evidence presented by the limited Chevrolet records existent, it seems likely that a similar standardization of Chevrolet chassis was put into place. For example, early contracts included chassis type NZ, for fire trucks, while on later contracts the fire trucks were built on NE chassis. It is possible that field lighting units, specifically J-5, were also built on NE chassis, but there is no direct evidence to support this. Dump trucks and tractors continued to use specialized chassis owing to their provision for overload springs.

E-5 Aerial Gunnery Trainer Assembly

One of the most unusual and today uncommon vehicles produced on the G-506 cab and chassis was the aerial gunnery trainer truck. The body for these trucks was made by the Herman Body Company of St. Louis, Mo. Whereas most 1½-ton 4x4 Chevrolets were assigned US Army registration numbers beginning with 3, indicating 1½-ton capacity, these trucks bore registration numbers beginning "60," the number used for Armored Cars and special technical vehicles.

While today many people use the designation "E-5" to refer to the truck, in fact that designation refers to the "E-5 Aerial Gunnery Trainer Assembly" which is mounted on the truck. The stock number of the unit was 5000-912000, and the specification number 94-24741.

The February 15, 1943 edition of Technical Order 11-65-1, which is specific to this trainer, states that the trainer comes with interchangeable mounting rings, one to fit the Sperry type A-1 turret, one for the Martin A-3 turret, and one for the Bendix A-4, all of which were upper turrets, as well as the Crocker-Wheeler A-8 training turret. Technical Order 11-65-12 mentions that the E-5 has also been used with Consolidated tail turrets, but that this "... requires considerable revision of the mounting, which, to date, has been made locally."

Attached to the front bumper were a pair of Joyce-Cridland 301RS ratchet jacks. Four such jacks were furnished with each truck. Cross beams just ahead of the rear wheels did the same, although those jacks were furnished with taller stand assemblies. These jacks were used to level the truck prior to firing the weapons. Operators were cautioned that a difference of 16-inches between front and rear wheels would result in the vehicle being on a 7-degree angle, and beyond that engine lubrication would be affected on the end connecting rods. Further, a water bypass shut off cock was installed, which was to be opened when the front of the truck was lowered to prevent steam pockets from forming.

Technical Order 11-65-12 notes that the E-5 could be fired while the truck was in motion, "...provided safety precautions are taken for stray rounds, stability and road surface."

A transmission-mounted power take off was used to drive a 24-volt type O-1 Leece-Neville 100-amp aircraft generator mounted under the front driver's side corner of the bed. The generator was not driven directly, but rather the PTO-shaft turned a line shaft supported by two ball-bearing pillow blocks. A pulley on the line

Above: *A cab-and-chassis version of the G-506 Chevrolet was initially designated the G-4113, and from 1942 on was called the G-7103. These had a 145-inch wheelbase and provided a platform for specialized bodies. Seen here is G-4113 registration number W-607642, which was built under contract W-398-QM-9402 and accepted in March 1941.* National Archives

Below: *This Chevrolet ton-and-a-half 4x4, registration number 3439666, was constructed as a cargo truck with winch, but appears here in the format of a cab-and-chassis truck, with no body installed. Although this particular vehicle is not known to have been one of the many G-506 Chevrolets that originally were intended as cargo trucks but were delivered as cab-and-chassis trucks, this vehicle is shown in this section because comparable contemporary walk-around series of photos of a true cab and chassis are not available.* General Motors LLC

Above: *G-506 registration number 3439666 is viewed from the left side. Unlike the authenticated cab-and-chassis G-506, registration number W-607642, previously shown, this vehicle has a spare tire carrier with a wheel but no tire to the rear of the left side of the cab.* General Motors LLC

Below: *Some of the special-purpose bodies that were installed on G-506 Chevrolet cab-and-body trucks were fire trucks and E-5 turret trainers. All of the 1945 model-year G-506 Chevrolet dump trucks were sourced from cab-and-chassis vehicles.* General Motors LLC

Without a body in place, details of the right rear spring hanger, the rear of the rear axle, and the bumperettes are more visible than usual. A holder for a five-gallon liquid container was fastened to the rear part of each running board. General Motors LLC

shaft is used to turn four V-belts, which then turn the O-1 generator. Above the generator is the voltage regulator and relay switch. These control the flow of electricity to two large batteries mounted in a metal box behind the truck cab. Because the truck engine drove the generator powering the trainer operation, and because gunnery training often took place in very hot regions, it is not uncommon to see wartime photos of the trucks with the hoods up in an effort to improve engine cooling.

A two-level frame is used to support the turret. Behind that frame is a stowage box used to store ammunition, tools, jack stands, and the canvas cover that protects the turret when not in use.

Unfortunately, few records related to the development, procurement and production of these vehicles have surfaced. However, a memo from the Headquarters, Army Air Force Chief of Staff on 26 August 1942 states. 'Though the Air Service Command had developed and was "about to procure in quantity" E-5 Gunnery Trainers....'

A second notation from the same organization includes the passage "In September, when E-5 trainers were in use...."

Together, these two comments lead one to believe that production began in August and ramped up quickly. However another report dealing with the procurement of aerial gunnery training equipment includes references to the E-5, as follows:

"Despite the uncertainty over the continued use of E-5 trainers, by the summer of 1942 the modified type was proving its efficacy, especially since its installation could accommodate the Martin and Bendix upper and the Sperry upper and lower turrets."

The second reference, citing that General Arnold had set 140,000 flexible gunnery graduates a year as the goal, stated that one E-5 was needed per 15 students, and that 1,226 to 1,800 of the trucks would be needed to achieve this, the final quantity being dependent on the duration established for the training course.

A variety of turret types were mounted on the trucks, ranging from the Crocker-Wheeler A-8 Training Turret to the Bendix, Martin, Sperry and Consolidated turrets found on US bombers during WWII. The A-8 was made by Crocker Wheeler Electric Manufacturing Company of Ampere, New Jersey; it was frequently armed with the Browning-designed M11 Remington 12-gauge semi-automatic shotgun. Owing to gun and ammunition shortages, the turrets installed on early E-5 Trainers were often armed with .30 caliber M1919 guns rather than the .50 caliber M2 guns used in combat zones. As students progressed, and as wartime shortages eased, the turrets and guns progressed as well, with Sperry Type A-1, Martin Type A-3, Bendix Type A-4 upper turrets being installed.

Truck, Trainer, Aerial Gunnery, Type E-9

The E-9 was an advance from the Type E-5 in that in it increased turret mounting options to include the Sperry A-2 lower turrets and Bendix A-16 chin turrets. The Type E-9 was produced by the Keystone Trailer and Equipment Company of Kansas City, Missouri on contract 2257-AC-1 beginning in July 1943, and completed in November of that year.

Unlike the E-5, the E-9 could only be fired when parked, and was to always be braced with the furnished jacks when firing. The leveling jacks were to be positioned on the four corners of the truck. Like the E-5, the E-9 featured a PTO-driven generator which supplied the electricity to operate the turret.

In addition to a PTO-driven generator, the truck itself was modified to use as an E-9 platform by relocating the fuel tank to a position in front of the large stowage box

In a rear view of a G-506 without a body installed, the emphasis in on the dual rear wheels, the two bumperettes, and the tow pintle between them. A wire-mesh screen is attached with brackets to the rear of the rear window, to protect the glass from damage. Partially visible below each of the bumperettes are the lever type shock absorber connecting links and axle mounts. General Motors LLC

Above: *The winch is partially visible above the front bumper of G-506 registration number 3439666. Set into a recess in the bottom of the bumper is the cable roller, which helped to guide the winch cable and allowed it to move freely through the obstruction posed by the bumper.* General Motors LLC

Below: *This G-506 Chevrolet, registration number 3485899, was one of 468 1944-model-year G-7107 cargo trucks from contract DA-W-3980QM-517 that were delivered without bodies, thus rendering them cab-and-chassis models.* General Motors LLC

Above: *The same vehicle, registration number 3485899, is seen from the front left. A spare wheel with no tire is mounted on the carrier on the chassis frame. No mud flaps are provided for the rear tires, and by this time, the Chevrolet maker's stamp had been eliminated from the hood.* General Motors LLC

Below: *G-506 Chevrolet cab and chassis registration number 3485899 had the liquid-container holders on the rear parts of the running boards. The tail pipe extends along the left side of the chassis frame.* General Motors LLC

In a right view of the same vehicle, registration number 3485899, the battery box is positioned to the immediate rear of the running board, and the bottom corner of the cab is cut out to accommodate the battery box. The two brackets for supporting the fuel tank are on the side of the frame above the rear drive shaft. General Motors LLC

at the rear of the truck. The stowage box also served as the mounting position for the spare tire.

To date, few photos have surfaced of the Type E-9 in use, which suggests that production of this variant was limited.

Adaptation of the E-5

The introduction of the Boeing B-29 Superfortress to the Army Air Force inventory late in the war brought with it the need for support equipment. The Second Air Force modified an E-5 trainer to become a auxiliary power unit to supply electricity to a B-29 during operations on the ground. To do this, the turret, generator and controls were removed from the E-5, and in its place a trio of 300 amp 28.5 volt generators installed, along with a cooling fan.

The unit-modified truck was supplied to Wright Field for a 300-hour evaluation toward a series of modifications of the trucks. A 25 July 1945 communications on this subject contains the passage "...received several months ago permission to modify fifty each of the now obsolete E-5 gunnery trainers...", and further states that a teletype had been sent on 23 February 1945 requesting "authority to increase the total number of modified E-5 trainer power units from fifty to seventy two."

Once the vehicle was in their possession, Wright Field engineers quickly determined that the power takeoff and drive line, originally intended to power a single 100-amp generator would be overburdened with a load 9 times that for which it was originally intended. Further, the cooling provision for the generators themselves was inadequate, and the report commented that this would surely lead to generator failure due to overheating.

For this reason, after lingering in storage for a year, the vehicle ultimately had the special equipment removed from it and was returned to the Wright Field motor pool.

Above: *A Chevy factory driver waits inside a nearly completed, early-model G-506 truck without a body while the man standing alongside the vehicle fills out a tag. Documents are taped to the right windshield. The bumper, the hood, and the fenders have a darker tone and more sheen than matte Olive Drab paint on the rest of the front end of the vehicle, evidence that the final coat of paint has not yet been applied. Interestingly, neither this truck nor the one shown below have tow hooks atop the bumper.* General Motors LLC

Below: *Newly painted and completed cab-and-chassis trucks are leaving the Chevrolet assembly line. At the bottom of the ramp, one of the vehicles is driving through a spraying booth with canvas windscreens on the sides. The purpose of this drive-through was to check the windshield seal for leaks.* General Motors LLC

Above: *As a means of training top-turret gunners, the U.S. Army Air Forces acquired a number of E-5 Turret Trainers. These consisted of a G-506 Chevrolet cab and chassis with a frame constructed on the rear for mounting the various types of dorsal turrets in use in the Air Forces. There was even an E-5 mount for a Sperry ball turret. The E-5 shown here is based on a 1942-model-year G-7103 cab and chassis, registration number 6061452. A Sperry A1-A dorsal turret of the type used on the B-17E and B-17F is installed on the frame. To the front of the truck, two men are adjusting screw jacks under the bumper, to stabilize the vehicle.* National Archives

Below: *This and the following two photographs of a Martin 250CD-5 power turret on a stand were found in a file of General Motors archival documents dealing with the E-5 Turret Trainers, and are presumed to show a turret that was to be installed on an E-5. This view is from the left rear of the turret. The Plexiglas dome has been removed, permitting a view of the gun-sight yoke and gun controls between the receivers of the machine guns. Below the turret ring is the back of the gunner's seat.* General Motors LLC

A man wearing a cap and headphones is sitting in the seat of the Martin 250CD-5 power turret. To the front of his head is the gunsight, mounted on a yoke that tilted upward in unison with the elevation of the guns. To the front of the sight are the feed mechanisms for the guns. General Motors LLC

In this view, the Plexiglas dome has been installed on the Martin turret. For training purposes, the dome had the virtue of protecting the gunner, the gun receivers, and the interior of the turret of the elements, but on the minus side, it made for a very hot location when the sun was shining. To alleviate the heat buildup, on the top of the dome is a wheel-shaped vent, also made of Plexiglas, which was twisted to open and close it. General Motors LLC

A top-turret gunner trainee is practicing on a Chevrolet E-5 Turret Trainer. He is using a special sighting device, apparently one with a periscope head on the front end and a padded eyepiece on the rear end. National Archives

Eight E-5 Turret Trainers are lined up at an unidentified location. None of the turrets have their Plexiglas domes on, and not all of the turrets are of the same type. From this elevated angle, details of the design of the top of the support frame for the turret are apparent.
General Motors LLC

Above: *Around early 1943, an E-5 Turret Trainer has been rigged as a so-called Spotlight Trainer at Tyndall Field, Florida. In the foreground is a spotlight, which was used to project a moving beam of light on the white wall in the background. Meanwhile, the student gunner in the turret would train the .50-caliber machine guns on the moving beam. A photo cell attached to the muzzle of the right-hand gun would register a hit on an automatic counter whenever the student "fired" the gun on-target.* National Archives

Below: *During a gunnery training exercise, eight E-5 Turret Trainers are arrayed in close formation. Note, the turret on the third vehicle is on a stand that is lower than those on the other vehicles. A narrow platform or catwalk was mounted on the front of the support frame, and on the first four vehicles, men are standing or crouching on these platforms. On the front bumpers were mounted two prominent wrap-around plates; jacks were placed under these plates to stabilize the vehicles during the firing of the guns.* General Motors LLC

Above: *One very famous graduate of the Flexible Gunnery School at Tyndall Air Force Base, Florida, was movie star and 1st Lieutenant Clark Gable, shown here trying out a Sperry A-1 dorsal turret with the dome removed on an E-5 at Tyndall, in or around January 1943. The construction of the top of the main frame for the turret is shown in close detail, including the turret ring for that specific turret, part number 2193-5, and several of the turret-ring brackets, part number 2193-2.* National Archives

Below: *Lieutenant Gable is adjusting himself in the Sperry A-1 turret on an E-5. Directly in front of Gable's face is the Sperry computing gun sight, which automatically calculated the amount of lead necessary to hit a flying target, and corrected for the fall of the bullets and ballistic deflection caused by wind.* National Archives

Above: *An E-5 equipped with a Martin A-3 power turret is viewed from the front. A Martin logo decal is on the shield below the top frame. In this turret, the .50-caliber machine guns were fed ammunition from their inboard sides. Booster motors with cogged wheels helped feed the ammunition to the receivers, and these wheels are visible in this photo.* National Archives

Below: *A shirtless driver is at the wheel of an E-5 Turret Trainer, while a gunner sits in the turret. The E-5 provided gunner trainees with valuable experience in learning how to aim and fire the guns at moving targets while at the same time under motion.* Taigh Ramey collection

Above: *The Chevrolet E-5 Turret Trainers lent themselves to a number of turrets in use by the Army Air Forces and the Navy. These vehicles at a desert firing range are equipped with Consolidated turrets of the type used in the B-24 Liberators.* Taigh Ramey collection

Below: *Top-turret gunners are learning the ropes on five E-5 Turret Trainers at an undisclosed location, date unknown. Several different types of turrets are in use. On the first two turrets, the left guns have the later-style barrel shroud with round perforations, while the right guns have the early-style barrel shrouds with elongated cooling slots. The second truck is registration number 6044532, while the third truck is 6044550.* National Archives

A staff sergeant crouching on an E-5 Turret Trainer points to something in the air while the man standing on the ground prepares to load .50-caliber ammunition. A man is standing inside the Sperry A1-A turret. This undated photo was taken at Blythe Field Army Airfield, Blythe, California. National Archives

A Martin amplidyne dorsal turret is mounted on the frame of a Chevrolet E-5 Turret Trainer at Mitchel Field, New York. Faintly visible near the center of the part of the Plexiglas dome of the turret that is visible from this angle is a transparent decal featuring a star in a circle over "Martin." National Archives

Above: *With two instructors supervising, a student is getting the feel of the operation of the Martin A-3 power turret on a Chevrolet E-5 at Laredo Army Air Force Base, Laredo, Texas, on July 18, 1943. Each type of turret used on the E-5 had its own specific turret ring. The Martin A-3 turret used turret ring part number 2193-3, along with the 2193-2 turret-ring brackets.* National Archives

Below: *In another photo of a training session on a Chevrolet E-5 at Laredo Army Air Force Base, the guns of the Martin A-3 are elevated, allowing a view of the right side of the main support: the curved structure to the right of the gunner's head. A similar support is on the opposite side of the turret, and gunners were instructed to grab the main support when hoisting themselves into the Martin turret. To the front of the gunner's face is the optical gun sight.* National Archives

At an unidentified training base, student gunners are practicing on E-5 Turret Trainers. The nearest truck has a Martin amplidyne dorsal turret, but the other trucks have another type or types of dorsal turrets. In the background, one of the gunners is loading ammunition into one of his .50-caliber machine guns. National Archives

Above: *Private Alfred C. Edwards, of Richmond, Virginia, is in the dome-less Martin turret on a Chevrolet E-5 Turret Trainer as his instructor monitors his progress. At this stage of his training, Private Edwards was trying to master the technique of leading a flying target by at least 40 feet.* National Archives

Below: *Top-turret gunner trainees on E-5 Turret Trainers are in the midst of live-fire practice against target sleeves being towed by small planes. Sanding on the platforms on the rears of the turret supports are the corporals and sergeants who are instructing the trainees. All four of the trucks are equipped with Martin amplidyne turrets.* National Archives

Above: *Chevrolet E-5 Turret Trainers are lined up as closely as possible at Harlingen Army Airfield, outside of Harlingen, Texas. The gunner trainees are firing at a mix of stationary and moving targets. The first four turrets are Martin amplidynes and are mounted on low stands. The rest of the turrets in the background are of a different type and are installed on taller stands; they are armed with Browning .30-caliber machine guns. The taller stands accommodated turrets with pedestals.* National Archives

Below: *On July 2, 1943, this E-5 Turret Trainer assigned to the 403rd Bombardment Squadron was photographed to illustrate a Fifth Air Force combat evaluation report. At the time, this squadron was serving in New Guinea. The turret is a Martin amplidyne type with Plexiglas dome installed. The 403rd Bomb Squadron modified its E-5 Turret Trainer by installing an intercom, so the instructor could communicate directly with the gunner while he was operating the turret.* National Archives

Above: *The 403rd Bombardment Squadron's E-5 Turret Trainer with intercom system installed is viewed from the left side with a gunner seated in the Martin turret and the instructor, standing by the driver's door, communicating instructions to the gunner via the intercom. Note the intercom control box to the side of the driver's seat. The E-5 Turret Trainers had a power unit, driven by the power takeoff, to supply electrical current to the turret, and a good view of the power unit, in the box along the side of the chassis frame to the rear of the cab, is available here. The louvered side door of the box containing the power unit is open.* National Archives

Below: *A Consolidated tail turret is mounted on this E-5 Turret Trainer, registration number 6044633, photographed on October 20, 1942. The copy of the photograph shown here originated from the Las Vegas Army Gunnery School at Las Vegas, Nevada, and presumably was taken there. The number 72 is stenciled on the side of the box on the rear of the body.* National Archives

The same E-5 Turret Trainer with a Consolidated tail turret, numbered 72, is viewed from the rear, with the turret traversed to the rear. This type of turret was used on Consolidated B-24 Liberator bombers, and the turret also was adapted as a nose turret on some B-24s. National Archives

Above: *The Consolidated tail turret is viewed close-up on an E-5 Turret Trainer at Las Vegas Army Gunnery School on October 20, 1942. Partially visible in the right background is another E-5 Turret Trainer, marked "81" on the side of the box on the rear of the body.* National Archives

Below: *A special, cantilevered support rig was devised for the E-5 (Modified) Turret Trainer to adapt a Sperry ball turret to the vehicle. This substantial support was installed on four legs, to support and hold the ball-turret assembly, including its support frame, trunnion ring support. The photo was taken at the Las Vegas Army Gunnery School on September 25, 1942.* National Archives

A Sperry ball turret mount is viewed from the left side on an E-5 (Mod.) Turret Trainer at the Las Vegas Army Gunnery School, September 25, 1942. When installed in a bomber such as the B-17 or the B-24, the ball turret could traverse 360 degrees, and it was possible to depress the turret to 90 degrees below the longitudinal centerline of the plane. On the E-5 (Mod.) mount, depression and traverse of the Sperry ball turret was limited. National Archives

Above: *A jeep with a pedestal-mounted .30-caliber machine gun is following a modified E-5 Turret Trainer at a firing range at Casper Army Air Field, Wyoming, during World War II. The E-5 evidently has been modified for the training of waist gunners, as the lower part of the turret support has been filled in with wooden slats, and a .50-caliber machine gun barrel is protruding through the opening above those slats. In the background between the jeep and the E-5 is the tail turret from a B-17 Flying Fortress. Machine guns are set up in several of the huts in the background, no doubt for the training of waist gunners.* National Archives

Below: *The Truck, Trainer, Aerial Gunnery, Type E-9 was the final development of the 1 ½-ton 4x4 Chevrolet-based trucks used for this purpose. The turret supporting structure differed considerably from that used on the E-5, allowing the mounting of belly and chin turrets without modification to the trainer.*

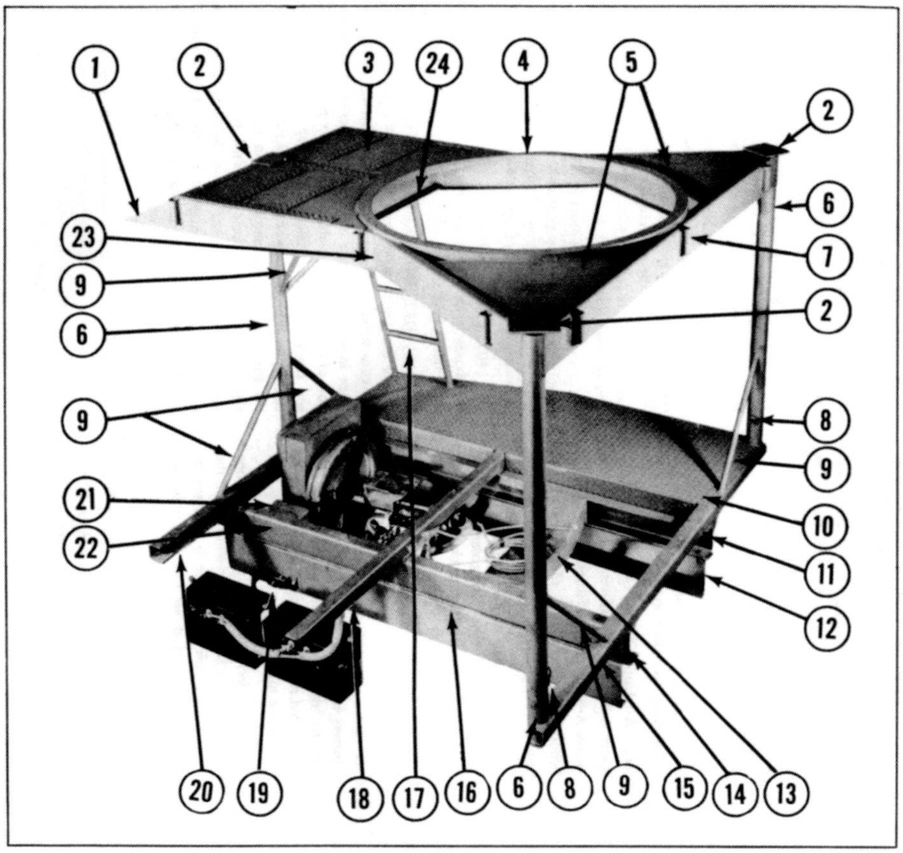

Index No. | *Nomenclature*

1. Support Channel, Front Left
2. Footplate, Upper Turret Support
3. Front Platform Assembly
4. Basic Mounting Ring and Mounting Ring for Sperry A-2 Turret
5. Floor Plate, Top Framing
6. Column, Vertical
7. Socket for Handrail
8. Loop for Tarpaulin Rope
9. Brace, Vertical Column
10. Running Board
11. Mounting Sill
12. Left Frame Rail of Truck
13. Lower Turret Support Mounting
14. Brace, Cross to Mounting Sill
15. Cross Sill, Rear
16. Cushion Sill
17. Ladder
18. Cross Sill Center
19. Belt Adjuster Bracket
20. Cross Sill, Front
21. Mounting Channel, Adjustable Bearing Housing
22. Filler for Item 21
23. Side Channel, Top Framing Assembly
24. Diagonal, Top Framing

Figure 8—Detail of Turret Supporting Structure with Power Unit and Air Blower in Place

Above: *Like the E-5, the E-9 used an aircraft generator, powered via PTO from the truck engine, to supply current to operate the electric turret. The turret was supported by this structure. Various mounting rings were supplied in order to accommodate the different types of turrets. The superstructure is shown here with the mounting ring for a Sperry A-2 lower turret.*

Below: *The chief advantage of the E-9 trainer was the ability to mount underslung turrets, like this Sperry A-2, without modification to the trainer. A fire interrupter limited the firing arc of the A-2 when installed on the E-9 to 75-degrees horizontal out of the rear of the truck.*

Above: *The E-9 superstructure came with removable railings around the upper platform. These protected training personnel when the A-2 turret was in place, but were removed when upper turrets were mounted, so as to not limit the field of fire. The large tool box stretching across the rear of the truck also housed the fuel tank.*

Below: *The Bendix A-16 remote chin turret could be installed on the E-9. Unlike the other turrets used with the E-9, this turret, commonly used on the B-17G, did not house the gunner.*

Above: *In addition to the various undermounted turrets, the E-9 trainer could accommodate various upper turrets, including the Sperry Type A-1, shown here, as well as the Martin Type A-3 and Bendix Type A-4 turrets.*

Below: *This E-9 Turret Trainer has been modified to a moving platform for training waist gunners. The support frame for a turret has been removed, and a raised platform with three pedestal-mounted .50-caliber machine guns and ammunition boxes has been installed. A Stinson L-5 Sentinel is buzzing two gunner trainees. Given the proximity of aircraft and buildings in the background, the gunners were carrying out a mock firing exercise as evidenced by the empty feed shoots.* National Archives

Above: *This E-9 Turret Trainer has been converted to support a section of B-29 Superfortress fuselage containing top and side blister windows for the gunners. The gunners controlled the dorsal and ventral turrets remotely, using gun sights aimed through the blister windows. Tagged onto the rear of the fuselage section is a rough mockup of the tail-gunner's compartment, with the gun sight visible through the windows. As no guns were present on this training device, it probably was intended for familiarizing the gunners with operating their sights and controls while under motion. For this practice session, a Stinson L-5 is buzzing the turret trainer.* National Archives

Below: *The Chevrolet E-9 could be adapted for manually operated machine guns, as seen in an April 22, 1945, photo of three student gunners behind .50-caliber machine guns. The weapons are on K-6 machine-gun mounts, which were used for enclosed waist-machine-gun positions. On these mounts, the gun barrel protruded through a slot in a can-shaped, swiveling fairing. Two different configurations of ammunition boxes, mounted high and low, are present.* National Archives

Above: *In the final months of World War II, the U.S. Army Air Forces saw a need for a mobile, external power plant for use in the ground maintenance of B-29 Superfortress heavy bombers. Specifically, the need was for a 900-amp, 28.5-volt DC power source with which to service and test components of the B-29s. To that end, by November 1944 the Army Air Forces was testing a power unit consisting of three USAAF Type R1 generators mounted on a converted E-5 turret trainer. One of those units is shown here; it was based on G-506 Chevrolet registration number 3108366.* National Archives

Below: *The same mobile power unit, registration number 2108366, is viewed from the right rear. The R1 generators were in a metal housing built around the frame that once had held a machine-gun turret; the upper part of that frame is visible above the housing. Stenciled in white on the rear of the generator housing and on the side of the box on the rear of the chassis is "MOBILE EXTERNAL POWER / PLANT FOR USE ON B-29." Portable fire extinguishers are on the rear of the vehicle and on the front fenders, and a spare tire was attached to the box on the rear of the truck.* National Archives

Above: *In a hangar at an unidentified airfield on 13 November 1944, a G-506 Chevrolet with a mobile external power plant is furnishing electrical power to a B-29 Superfortress. The power cable may be seen issuing from the left side of the vehicle and up to the right side of the bomber's fuselage. This vehicle has the same unit markings on the front bumper as the truck in the two preceding photos, and a placard reading "ROAD TEST" is attached to the left tow hook.* National Archives

Below: *The following series of photos, dated 29 November 1944, was intended to document the steps required to convert an E-5 turret trainer to a mobile power plant. This photo, showing the area within the turret support frame, includes a placard listing the procedures in step #1: remove turret and controls, floor boards, gas tank, battery box, and relay box.* National Archives

Above: *Step #3 in the conversion was to remove the generator and drive assembly in the box with the louvered door to the rear of the left running board. This generator had powered the turret on the E-5 turret trainer. Between steps 1 and 3, a modified gas-tank mount had been installed just above the bottom of the former turret support frame.* National Archives

Below: *Skipping ahead, step #16 is shown completed in this photo: installation of the control panel, with all electrical connections completed. There are separate controls and gauges for the three Type R1 generators.* National Archives

The original generator and drive assembly for the E-5 turret trainer have been removed and are lying on the floor below the left side of the chassis frame. In subsequent steps, the three Type R1 generators would be prepared with their drive pulleys and sheaves, and these, along with a cooling fan, would be installed in the space within the former support frame. National Archives

Above: *Step 25 in the conversion was mounting the rear fire extinguishers. Tests of the vehicle indicated that the cooling fan of the generator wasn't sufficient for sustained operation of the power plant, and the truck's engine and power takeoff were not deemed robust enough to transmit a power output of 900 amps at 28.5 volts DC for a sustained length of time.* National Archives

Below: *Step 26 in the conversion of the mobile power plant is documented here: the mounting of a fire extinguisher on each front fender. The U.S. Second Air Force was authorized sometime before May 1945 to order convert 50 E-5 turret trainers to mobile external power sources, but it is not clear how many of these vehicles were converted.* National Archives

Chapter 8

Bomb Service Trucks

The Chevrolet Bomb Service truck, M6, was the result of the Army Air Corps desire to improve the 4x2 M1 Bomb Service truck produced by Ford and others. The function of the bomb service truck was to load, unload and tow bomb trailers transporting bombs up to 1,000 pounds in size from their storage point to the aircraft. The April 1941 list of items they felt should be improved over the M1 included:

(1) Four-wheel drive
(2) 1½-ton chassis
(3) Four speed transmission
(4) Two speed transfer case
(5) Short wheel base
(6) Single tires, front & rear
(7) Short turning radius
(8) Low silhouette
(9) Additional seating capacity for three men in the rear
(10) Removable or collapsible top over the cab.

Several manufacturers submitted designs and bids with the development contract finally going to the Chevrolet Division of General Motors Corporation.

Chevrolet based their submission on the standard 1½ truck, albeit with extensive modifications in order to meet the requirements laid out above. Their efforts paid off, and every one of the requirements were met.

On 10 July 1941, Ordnance Committee Minutes (OCM) 16969 recommended the Chevrolet submission be standardized as the "Bomb Service Truck M6." This recommendation was approved by OCM 17116 on 11 August 1941, with the same action relegating the M1 as Substitute Standard.

It is worthwhile to point out that unlike most of the vehicles in this book, which were developed and initially procured under the auspices of the Quartermaster Corps, the bomb service truck was an Ordnance Department procurement, and thus had its own SNL "G-" number, G-85, whereas the bulk of the other vehicles in this volume are SNL G-506 – the 500-series numbers reflecting items taken over by Ordnance from Quartermasters in July 1942.

The Army, anticipating the August Standardization, had begun negotiations with Chevrolet for the production of these vehicles, soliciting a bid on 20 May 1941, which the company responded to a week later. A contract for the production of 491 of these trucks, number W-374-ORD-1131, was issued on 29 September, and production began the following month. During October 68 of the trucks were delivered, believed to have started with serial number 14NQ10-1684, followed by 185 in November and 238 in December. These trucks were assigned registration numbers 004373 through 004863.

Contract W-374-ORD-1185, production order T-2068, was also issued in September 1941, actually predating contract 1131 by a couple weeks, but production of these 116 vehicles did not begin until January 1942. A supplement to this contract, production order T-2276, covered 127 additional trucks which were delivered in February 1942. All of the trucks on contract W-374-ORD-1185 appear to have been built in Baltimore, with 148 being completed in January and 95 in February. The initial group bore registration numbers 006070 through 006185, while those on the supplement were assigned registration numbers 006186 through 006312.

Even more bomb trucks were ordered, and also produced in Baltimore, on per W-374-ORD-1281, which were contracted for on 15 January 1942. Production of these 435 trucks began in April of that year, with 128 being delivered, followed by 239 in May, 63 in June and the final five in July. The registration numbers for these vehicles ranged from 007987 through 008421.

Production of the M6 continued to accelerate, with a further 548 being ordered on contract W-374-ORD-1580, executed 15 July 1942. The first 206 were delivered in November, followed by 325 in December, and 17 in January 1943. The trucks were assigned registration numbers 0012010 through 0012555. A single truck, registration number 0012557, was ordered on a supplement covered by production order T-3952.

Two contracts issued in August 1942 (W-374-ORD-2216) and December 1942 (W-374-ORD-4052) truly

Above: *Based on the chassis and modified cab of the G-506 Chevrolet 1 1/2-ton 4x4 truck, the M6 Bomb Service Truck was standardized on 10 July 1941 and was tasked with loading bombs onto a bomb trailer and then towing the trailer to the flightline, where the bombs were loaded into bombers. The M6 had a steel crane over the body of the vehicle, with a 2,000-pound-capacity bomb winch on the crane. The M6 Bomb Service Truck was designed to operate with the M5 Bomb Trailer, designed to carry bombs up to 1,000 pounds in size.* General Motors LLC

Below: *A bench seat for three men was in the rear of the M6, under the crane; the seated men faced to the right. Storage boxes were on the left running board and behind the cab. The cab was open, with a bow to the rear of the front seats to support a canvas cover. The hand-operated hoist, seen here at the bend at the front of the crane, was designed to slide along the rail of the crane, as desired. An arched support was toward the rear of the crane.* General Motors LLC

The cab of a 1942-production M6 Bomb Service Truck manufactured under contract W-374-ORD-1131 is illustrated, with the steering wheel removed for clarity. The dash was of a much different design than those of the other G-506 trucks, and featured military-type gauges. Data plates were attached to the glove box and the right side of the dash. The layout of the shifting levers and foot pedals was similar to the other G-506 trucks. General Motors LLC

brought the M6 into mass production, with the orders being for 1,701 and 2,359 trucks respectively. These two orders, along with a third order described below, meant that there would be M6s coming off a Chevrolet assembly line every month in 1943. The trucks on contract -2216 carried registration numbers 0022800 through 0024500, while those on contract -4052 bore numbers 0036252 through 0038610.

The final production contract for the M6 was W-374-ORD-5151, which was issued 28 July 1943. A total of 1,947 bomb service trucks were built on this contract, production of which began in October 1943. Production ended with 607 trucks being delivered in January 1944, and the final M6 deliveries of 364 in February 1944. The trucks on this contract bore registration numbers 0061361 through 0063307.

The total of 7,724 M6 bomb service trucks were produced in three plants, with Norwood building 5,092 examples, St. Louis producing 1,320 and Baltimore 1,311.

While Chevrolet assembled the trucks, the bomb service bodies were produced by the American Coach and Body Company of Cleveland. There were four types of bomb service bodies produced, the chief difference being in the number and width of the heavy oak planks that comprise the floor of the bed of the truck. The body variations were designated by design number, which were numbered one through four. There was also variation in the winch. Originally a Beebee Brothers winch was used. When they were unable to keep pace with demand, a Braden winch was used. The Braden was found to be unsatisfactory, and it was replaced by a J.H. Holan winch. James H. Holan was the president of American Coach and Body Company, and the son of the founder. Two types of Holan winch were used on the bomb service trucks, part number 5149-39 and part 5320. The Holan units were interchangeable as complete units, but the internal components were different.

These trucks were furnished with M1 Bomb Lift Trucks built by Weaver Manufacturing of Springfield, Illinois. The M1 was carried in the bed of the truck beneath the crane structure. This is a small, manual operated cart, and is not to be confused with the large M5 bomb trailers often seen under tow by M6 bomb trucks.

All the M6 Bomb Service Trucks were "1942 year model" trucks, and thus carry the Master Book and serial

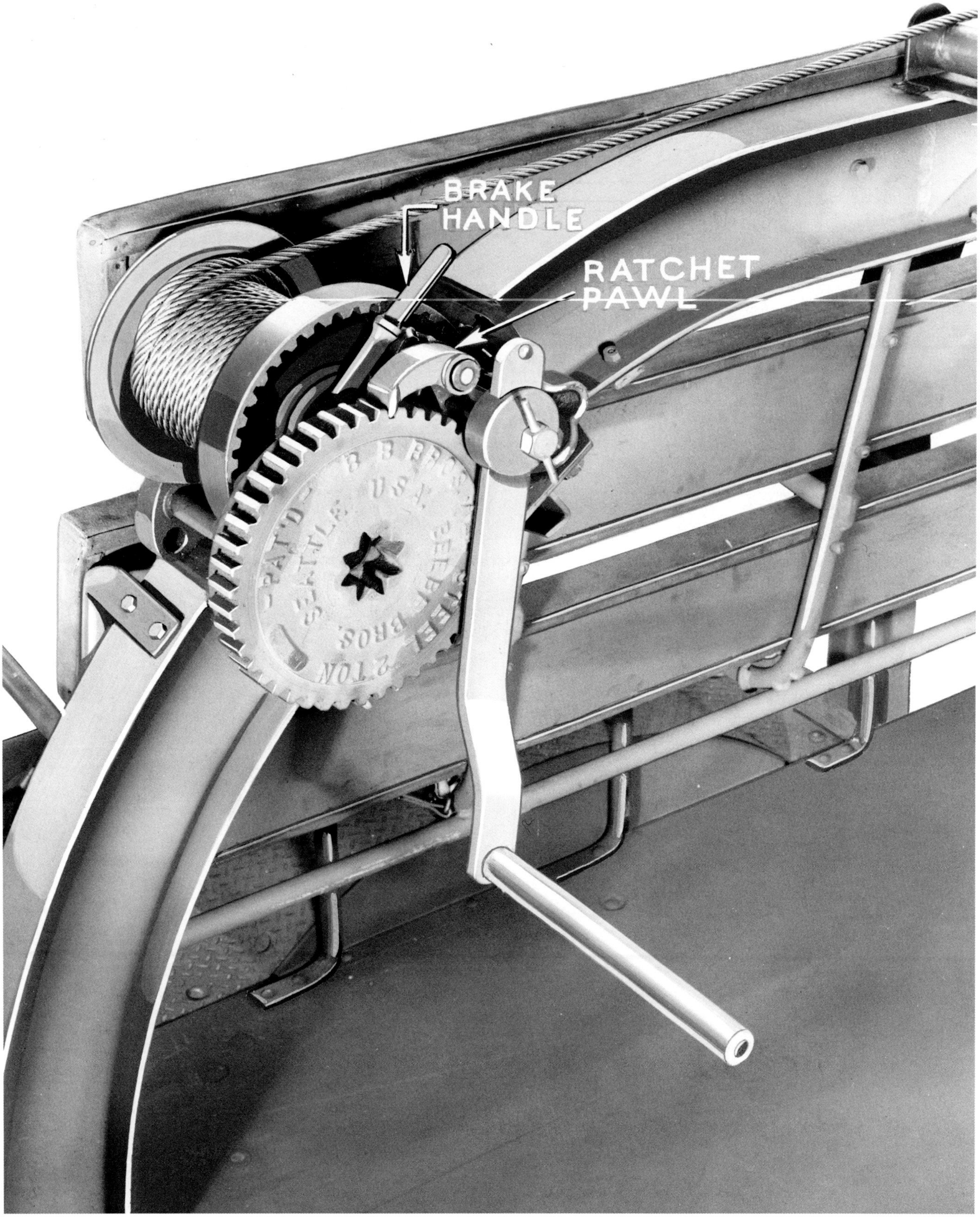

The two-ton winch on an M6 Bomb Service Truck is viewed from the left side. The winch was manufactured by Beebe Brothers of Seattle, Washington. The hand crank is on the left side of the winch and the cable drum is on the right; indicated in the photo are the brake handle and the ratchet pawl. General Motors LLC

The M5 Bomb Trailer was used in conjunction with the M6 Bomb Service Truck. Running front-to-rear on the trailer bed were rails, to keep the bombs from rolling. The dual front caster wheel allowed the trailer to be turned in its own tracks. A storage box on the front of the trailer bed had mountings for bomb fins, which, because of their fragile nature, generally were installed on the bombs at the time the bombs were being loaded in an aircraft. On the side of the trailer is the registration number, W-05193. General Motors LLC

number prefix of NQ, and the Chevrolet model designation G-7128.

It is noted in Engineering of Transport Vehicles, by the Chief of Ordnance – Detroit, that "On 10 July 1943 the Chevrolet Motor Division were requested to incorporate the spare tire assembly and suitable carrier. This change was released for production in 1943." However, truck registration number 0038610, presumably the last truck on the next to last M6 order, and presumably the last M6 made in 1943, is extraordinarily well documented photographically, as this truck was used at Aberdeen for A-Frame lifting device experiments and then later became the pilot for the air transportable bomb truck shown later in this book. No sign of a spare tire carrier is evident. Likewise, registration number 0062378, a truck from the last contract, was photographed in as-new condition at the Studebaker Proving Ground on 19 January 1944, and it too shows no sign of a spare. Truck 0062495 was photographed at Aberdeen Proving Ground in stock configuration, and once again no spare is evident.

Truck 0062512 is the only truck found in several years research that is clearly equipped with a spare tire, and that truck was modified to accept a cross country tire kit. The spare, which is mounted on the left side of the rear crane support, is of the type included in the kit, so it is uncertain if the spare was installed as part of the cross-country conversion or if the carrier was factory installed. No reference to bomb truck spare tire mounting is found in the ORD SNL G-85 Vol IV and G-506, nor any other vintage photos, leading us to wonder if this change was ever actually implemented.

The fuel system of the bomb service truck is unique among the Chevrolet 1½-ton 4x4 in that it has two fuel tanks, a 30-gallon main tank suspended from the frame rails and an 18-gallon auxiliary tank beneath the seat.

In addition to the 16 March 1944 recommendation of development of an air transportable version of the M6,

Above: *This 1943-model M6 Bomb Service Truck, registration number 0012244, was produced under contract W-374-ORD-1580. The blackout marker lamps are the type with the rectangular aperture. On this vehicle, the horn is mounted above the left fender.* National Archives

Below: *In this view of the left side of M6 registration number 0012244, a retainer chain is visible on the padlock on the storage box below the driver's seat. Stored on the bed is an M1 Bomb Lift Truck. The M6 had a short wheelbase: 125 inches.* General Motors LLC

The same M6 Bomb Service Truck, registration number 0012244, is seen from the front. The brush guard is the late type for G-506 trucks, with individual, welded vertical rods. No blackout headlight was installed. A large wingnut is adjacent to each bottom corner of the folding windshield. General Motors LLC

Above: *The rear crew bench seat, as seen on M6 0012244, faced to the right and accommodated three men. There were three separate back rests: one on the crane, with a separate cushion on the upper part of the rear support of the crane, and two below the crane. The fuel filler was on the right side of the vehicle, routed through an opening in the right front seat riser.* General Motors LLC

Below: *Unlike the other G-506 Chevrolet 1 1/2-ton 4x4 trucks, the M6 Bomb Service Truck had single rear wheels. The cab cover was a simple square of canvas, draped over a bow, with a window flap on the rear. The rear support arch of the crane was fashioned from steel I-beam. Whereas other versions of the G-506 Chevrolet 1 1/2-ton 4x4 truck had three-spoked steering wheels, the M6 Bomb Service Truck had a four-spoked steering wheel.* General Motors LLC

A tow pintle for hitching a bomb trailer was mounted below the rear cross member of the chassis frame. On the cross member were an electrical connection for the trailer and two eye bolts. To the sides of the cross member were tail light assemblies and the rear spring mounts. Sitting on the truck body is an M1 Bomb Lift Truck. General Motors LLC

Above: *During 1942, a winterization kit was developed for the M6 Bomb Service Truck, to better enable it to function in cold weather. The kit included a tarpaulin, which protected the front end of the vehicle and kept cold air from infiltrating the battery and engine areas from underneath the vehicle. The tarpaulin is shown here installed on an M6 at the Chevrolet Motors Engineering Division on 7 October 1942.* General Motors LLC

Below: *The M6 Bomb Service Truck with the winterization tarpaulin installed is viewed from the right side on 7 October 1942. In addition to the tarpaulin, the winterization kit for the M6 included an insulated battery box with a heating coil and gasoline heater; a primer system; an engine-oil dilution system; crankcase ventilation system; radiator cover; hot-water foot warmer; electric windshield defroster; 162-degree thermostat; fuel-pump diaphragm; and underchassis heater.* General Motors LLC

Above: *As seen from the right rear, the tarpaulin enclosed the sides of the cab but not the top and rear of the cab. An access flap secured with snaps was on each side of the tarpaulin below the floor level of the cab.* General Motors LLC

Below: *This photograph was part of the series taken to document the winterization kit for the M6 Bomb Service Truck, showing the right side of the engine compartment, and including in the center foreground the oil diluter, one of the heating components of the kit.* General Motors LLC

discussed in Chapter 13 of this volume, the Ordnance Committee had a few other actions related to the M6.

Among these, the 6 May 1943 Standardization of winterization equipment by OCM 20369, the 29 June 1944 Standardization of a cross country kit, including larger tires, for these trucks, and the 29 March 1945 OCM 27144 approval of development of a field-installed front winch kit.

However, even though production of the vehicles had ended in February 1944, the M6 would come to the attention of those developing military vehicles several more times.

In January 1945, the Air Ordnance Officer of the 9th Air Force reported to the Development Division, Chief of Ordnance, Detroit, that the need existed for a bomb service truck with a swinging boom. A swinging boom would allow the bomb service truck to operate in more confined areas than possible with either the M27B1 or M6.

The desired military characteristics of the swinging boom bomb service truck were:

(1) Equipment mounted on standard Ordnance Chassis, preferably Truck, 2 ½-ton, 6x6.
(2) Crane to be of swinging type capable of lifting four 500 lb. bombs or one 4000 lb. bomb at maximum radius without use of support legs.
(3) A body to be incorporated, if possible, to provide carryings space in vehicle for bombs."

Before this project had advanced very far, it was dropped as Ordnance learned that Headquarters, USSAFE had developed a modification of the Truck, Bomb Service, M6 which included a swinging boom. A contemporary report on the subject explains, "....the fixed overhead

In a view of the underside of the chassis of an M6 Bomb Service Truck taken for the Chevrolet Motors Engineering Division to document the winterization kit on 7 October 1942, the hoses for the heating coil for the battery are seen entering the bottom of the battery box toward the bottom right. Also in view are the bottom of the right running board and its supports (at left) and the two forward drive shafts: one for the front axle (longer) and one between the transmission and transfer case (shorter). General Motors LLC

Above: *Resting on a trailer hitched to Chevrolet M6 Bomb Service Truck registration number 0024296 is a K-3 Bomb Lift Truck with a large bomb mounted on it. This bomb lift truck was self-propelled, with two sets of caterpillar tracks on the front and two wheels on the rear. This truck was designed to position large bombs, up to 4,000 pounds, under the bomb bay of a bomber aircraft and lift the bomb up to its shackles.* National Museum of the United States Air Force

Below: *Chevrolet M6 Bomb Service Truck registration number 0038610, a 1943-model vehicle manufactured under contract W-374-ORD-4052, is shown at Aberdeen Proving Ground 22 March 1949. An A-frame boom has been fitted on the front end, apparently to facilitate lifting bombs that the truck could not drive right up to. This vehicle also has a front winch: a feature that was installed on existing and in-production M6s beginning in 1945. This winch had a 10,000-pound line capacity. Ordnance Museum*

Above: *A Mk. 6 Mod. 1 Bomb Truck with a general-purpose bomb poised on it is hitched to M6 Bomb Service Truck registration number 0036417, produced under contract W-374-ORD-5052 and accepted in 1943. The Mk. 6 Mod. 1 Bomb Truck, as the sign says, was designed "to lift munitions and fuel tanks to carrying position on fighter aircraft." Note the caution stripes painted on the rear of the crane.* National Museum of the United States Air Force

Below: *In an undated photograph, an M6 Bomb Service Truck has an M5 Bomb Trailer hitched to it. A large bomb and its separate fin are loaded on the trailer. The M5 Bomb Trailer weighed 2,000 pounds and had a carrying capacity of 5,000 pounds.* Chemical Corps History Office

Above: *This M6 Bomb Service Truck, registration number 0062378, and the M6s in the following two photos were accepted in 1943 and were produced at Chevrolet's Detroit plant under contract W-374-ORD-5151. The code "E-2-30" is stenciled in white on the battery box. The photo was taken on 19 January 1944 for the Ordnance Operation, Studebaker Proving Ground.* TACOM LCMC History Office

Below: *M6 Bomb Service Truck registration number 0062495 was photographed at Aberdeen Proving Ground on 1 September 1944 during a test of cross-country tire kits. An M1 Bomb Lift Truck is on the bed of the M6. Note the cable roller on the top of the crane between the winch and the cable sheave at the end of the crane; this roller was to keep the winch cable from scraping on the top of the crane.* National Archives

In a photograph taken for the Office, Chief of Ordnance, Detroit, on 1 September 1944, M6 Bomb Service Truck 0062512 is depicted with cross-country tire kits installed. The cross-country tire kit was developed in 1944 to help the M6 maneuver over adverse terrain, such as that encountered at bomb dumps in undeveloped areas. Note the Olive Drab paint overspray on the light-colored seatback. National Archives

hoist and hand operated winch were replaced by a revolving pedestal mount and boom copied after the tractor crane, M1. Power was taken from the vehicle's power train and transmitted through chains, sprockets and transmission parts to three winches, providing full power to the lift line, boom shroud line and traverse. Outriggers and dual rear wheels were used to provide stability. The satisfactory bomb-handling performance of this improvised equipment indicated basic suitability of the modification."

Headquarters, Army Air Forces, requested that a modification kit be developed for the M6 which would permit the conversion of the trucks, and further requested that two pilot models be produced.

On 22 March 1945, OCM 27034 recommended that such a vehicle be developed, however preliminary discussion with industry leaders revealed that the requested 360-degree traverse would take a long lead time to develop, perhaps as much as five years. If rotation could be limited to 220 degrees, the vehicle could be developed rapidly by modifying existing designs.

It was decided to pursue development in two phases. The first phase was the procurement of two 220-degree swing mechanical cranes from Gar Wood, Inc. The second phase would be the purchase of two models of 360-degree swing cranes, two each from Trackson Company, Milwaukee, which would also be mechanical, and two from Milwaukee Hydraulics Corporation, which would be hydraulically actuated. One of each of these pilots was to be a light weight model.

The model designation for these various kits were listed in OCM 27703, and were:

(1) T34 mechanically operated, Gar Wood Industries
(2) T34E1 mechanically operated, Trackson Company
(3) T34E2 mechanically operated, Trackson Company, lightweight
(4) T34E3 hydraulically operated, Milwaukee Hydraulics
(5) T34E3 hydraulically operated, Milwaukee Hydraulics, lightweight

The first pilot of each model was shipped to Wright Field, Ohio, in July 1945. A second T34E2 was soon procured and sent to Manila, Philippines, for testing.

During the Korean War, the Office, Chief of Ordnance, at Aberdeen Proving Ground devised and tested the prototype Truck, Wrecker, 1 1/2-ton, 4x4, XM247. Converted from Chevrolet M6 Bomb Service Truck registration number 0012342, the XM247 was intended to serve as a light wrecker for recovering and handling administrative-type vehicles assigned to camp, post, or station motor pools. The bomb-handling equipment was removed, and a Code-A Model 535 Wrecker Kit was installed for test purposes. The tests of the vehicle at Aberdeen were conducted from 15 August to 30 September 1951. National Archives

In addition to the kit sent to Manila, kits T34, T34E1 and T34E3 were installed on three M6 bomb trucks for testing. Dual rear wheels were installed on all three trucks for stability. While all three did function, the conclusion of the test was that the M6 was too light and too unstable for use with a swinging boom. However, it was noted that the T34E3 with its hydraulic crane was the best all around unit.

The three lightweight pilots were sent to the Air Force Proving Ground, Eglin Field, Florida for testing, but this was never completed. The nomenclature of the units was subsequently changed to T49, with the various suffixes as above.

OCM 32620 of 27 January 1949 classified the M6 as obsolete, to be replaced by the M27B1 2½-ton 6x6 bomb service truck. However, on 21 June 1949, the Headquarters, US Air Force requested that the classification be changed yet again, from "Obsolete" to "Limited Standard." In doing so, they wrote "Since this item was declared obsolete, many requests have been initiated by fighter wings world-wide requesting that action be taken to continue issue of this item pending the design and development of an item more suited to fighter aircraft requirements. From all indications, the item 'Truck, bomb service, M27B1' does not meet operational requirements for fighter squadrons inasmuch as this item is too bulky, not permitting easy access to revetments or hardstands near aircraft. Armament personnel voice fear of injury to aircraft. Fighter aircraft normally carry five hundred or one thousand

Above: *The XM247 Wrecker Truck was equipped with double swiveling booms with a 6-ton lifting capacity, and dual power-driven winches with 15-tons capacity, all of which were located at the front of the wrecker body. A jack is installed below the front of the body to stabilize the vehicle during boom operations. Extending from the rear of the wrecker body is the towing cradle.* National Archives

Below: *The XM247 is viewed from above on 22 August 1951, showing the double swiveling booms in their travel position. After initial testing of the vehicle at Aberdeen Proving Ground, the XM247 served at the post's motor pool for further assessment. Motor pool reports revealed that the vehicle was very useful in daily service and had been able, during a 30-day period, to handle over 75 percent of the motor pool's wrecker requirements. This included being able to successfully handle an M34 2 1/2-ton 6x6 Cargo Truck.* National Archives

pound bombs and therefore do not necessitate the use of a vehicle equipped to handle up to four thousand pound bombs."

The Air Force's request brought about, at least for a while, the disposal of the then-existing inventory of M6 Bomb Service Trucks per OCM 32986 of 8 September 1949. Interestingly, the OCM also noted that both the conventional and airborne models were not distinguished from one another in stock records, and both were to be retained.

However, by 1951 the Air Force had declared some 1,700 M6 Bomb Service Trucks excess of their needs. In an effort to reutilize these trucks, the Office, Chief of Ordnance initiated an effort to develop an administrative-type wrecker on the surplus chassis in August of that year.

The resultant vehicle, designated XM247, consisted of a M6 chassis with a Holmes Model 525 wrecker bed installed. The dual boom, dual winch bed was driven via PTO from the truck. It was furnished with a tow bar, towing cradle, ground anchors, steering wheel clamp and other standard towing gear.

The modified vehicle was assigned to the post motor pool at Aberdeen Proving Ground for a 30-day evaluation. During that time, motor pool staff estimated that the XM247 handled 80% of the motor pool requirements.

In an effort to test the capacity of the equipment, personnel lifted a loaded 2½-ton 6x6 M47 dump

As seen in this right-rear view of the XM247 on 22 August 1951, the towing cradle is swiveled forward to its travel position. The report of the testing of the vehicle at Aberdeen argued that, notwithstanding the vehicle's many virtues, the open cab was undesirable for purposes of a wrecker, from a crew-comfort standpoint. National Archives

An extra-large front bumper was installed on the XM247, replacing the original bumper, which was much shorter in height. This enlarged bumper was better suited to the occasional need to push mired or disabled vehicles. The brush guard was the early, symmetrical type. National Archives

The XM247 Wrecker Truck is viewed from the rear at Aberdeen Proving Ground on 22 August 1951, with the double booms in the travel position. Test-report recommendations included widening the body by six inches, to provide additional wheel clearance. In addition, despite the substantial appearance of the rear cross member of the wrecker body, the testing staff felt that an even heavier-duty cross member was desirable, to provide better support for the towing cradle pivot bracket, which is seen here at the center of the cross member. National Archives

A few days after the end of the six-week testing period of the XM247 Wrecker at Aberdeen Proving Ground, the vehicle was photographed after hoisting an M43 3/4-ton 4x4 Ambulance and attaching a standard V-type tow bar, on 4 October 1951. Despite this vehicle's excellent performance during the tests at Aberdeen and the availability of many surplus M6 Bomb Service Trucks for possible conversion to the wreckers, this project proceeded no further than the prototype. National Archives

truck with the experimental wrecker, and successfully towed the dump truck away. The personnel did note that they believed this was the limit of the XM247's abilities.

The only real deficiency noted for the vehicle was the open cab, which the test report noted "No provision has been made to protect the driver from the weather. This could be very uncomfortable, especially during the winter season."

While the report concluded with a recommendation that such conversions be adopted for motor pool use, it appears that this recommendation went unheeded.

Chapter 9

Fire And Crash Trucks

Many types of fire-fighting vehicles were built on the G-506 chassis, even including cab over engine models for the US Navy. The bulk of the G-506-based fire trucks were built for the Army, however. In the years leading up to WWII, fire trucks were the responsibility of the Quartermaster Corps, and the Quartermasters established a system of classification for fire apparatus. When responsibility for fire trucks was shifted to the Corps of Engineers, they continued to utilize the same system.

This system assigns "Class" numbers to the very types of fire engines. These are three digit numbers, the first digit of which represents EITHER a truck's function OR its pump capacity. The 100-series trucks are airfield/crash trucks, the 300-series trucks have a pump capacity of 300 gallons per minute (GPM), and the 500-series trucks pump 500 gallons per minute, and so on. None of the G-506-based trucks had a pump capacity exceeding 500 gallons per minute.

The next two digits distinguish the various models of truck, within a type or pump capacity range, from each other.

The Class 110

The first of the conventional cab models were the Class 110 trucks. This class began with 30 vehicles requested on

The first of the Class 110 Airfield Crash Trucks, registration number W-5072, was photographed on 15 May 1941. On the side panel of the hood is a placard that reads, "MODEL 110, QMC." Chrome hand rails are attached to the upper parts of the sides and front of the crash body. A dark-colored non-slip panel, likely made of rubber, is on top of the rear fender. A clawbar door opener is on holders above the rear fender. General Motors LLC

In a view of the right side of Class 110 Airfield Crash Truck W-5072, two separate doors for storage compartments are toward the front of the crash body, and a pike pole is stored on holders above the clawbar, near the top of the crash body. Two portable fire extinguishers are secured to holders on the rear running board. General Motors LLC

an Invitation For Bid dated 30 October 1940, and subsequently ordered 13 January 1941 on W-398-QM-9250. The trucks were delivered to the Quartermaster Corps by Chevrolet Baltimore as cab and chassis beginning in April 1940, and an apparatus bed was installed by the Holabird Quartermaster Depot, creating an airfield crash truck. Deliveries of completed trucks began in June 1940.

These thirty trucks were equipped with a 100 gallon per minute capacity Hale pump. The pump was fed by a 250-gallon water tank housed in the apparatus bed. A firefighting foaming agent was premixed with the water, and the solution was discharged through two pre-connected 100-foot 1½-inch hoses, which were stowed in compartments on either side of the water tank.

At the rear of the truck was a reel with 100-feet of ½-inch high-pressure line with a carbon dioxide discharge horn. This was supplied by four fifty-pound CO2 cylinders mounted transversely in the apparatus bed, just behind the cab.

The initial 30 trucks, which bore Army registration numbers 50702 through 50731, were soon joined by more. The Quartermasters assembled 157 more examples at the Holabird and Memphis (TN) Quartermaster depots. These trucks, which carried registration numbers 501274 through 501428, featured American LaFrance pumps rather than the Hale pump of the previous groups, but the specifications were the same. They were built on G-7113NZ chassis ordered on contract W-398-QM-10328 in June 1941.

A further 40 examples of these trucks, which became designated Class 110, were built by Oren Fire Apparatus of Roanoke, Virginia. These trucks were assigned Army registration numbers 503961 through 504000. These trucks were assembled on G-7103NE chassis produced by Chevrolet's Norwood plant on contract W-398-QM-13532.

In accordance with War Department Circular 248, of 4 December 1941, the Corps of Engineers took over responsibility for firefighting from the Quartermaster Corps and assumed the responsibility for firefighting equipment procurement. Thus, the Engineers had built 150 more Chevrolet-based Class 110 trucks built, with registration numbers 503769 through 503918. These trucks were also constructed on G-7103NZ chassis produced by Chevrolet's Norwood plant on contract W-398-QM-13532.

Above: *The same Class 110 Airfield Crash Truck is viewed from the rear. A reel holding 100 feet of 1 1/2-inch high-pressure hose is on the center rear of the body. Stowed in compartments on the sides and attached to the Y-shaped dual connections on the lower left of the body are two 1 1/2-inch fire hoses. The connectors have valve handles on top. Two fire extinguishers are stored on each side of the rear running board.* General Motors LLC

Above: *Stored on holders toward the rear of the left side of the crash body are a fireman's axe and a crowbar. Chrome protectors are on the rear corners of the rear running board.* General Motors LLC

Above: *Above the storage-compartment doors toward the front of the crash body is a gauge. Below the doors is a chromed pump-flushing connection. The latches on the storage-compartment doors and the Mars siren on the front fender also are chrome.* National Archives

Below: *Class 110 Airfield Crash Truck registration number W-50702 is seen from the front. The Mars maker's mark is visible on the louvered front cap of the siren. Note the absence of blackout marker lamps on the front fenders.* General Motors LLC

While most of the Class 110 trucks were used in the Continental United States, 30 examples were supplied to Great Britain under Lend-Lease. Following the creation of the United States Air Force on 18 September, 1947 the Class 110 was redesignated the 0-2 by the new service.

On 21 April 1949, the Air Force decided to retain their Class 110 trucks and dispose instead of Ford 4x2-based Class 135 trucks, saying that the latter were more difficult to maintain and obtain parts for.

Class 135

The second type of airfield crash truck to be produced on the G-506 chassis was the Class 135. The Class 135, like the Class 110, was equipped with a high pressure, low volume pump. While in US use, the Class 110 was retained stateside, the Class 135 was widely used overseas. The official nomenclature describing these trucks was "Truck, Fire, Powered, Crash, Class 135, 4x4, High Pressure Fog Foam." The Table of Organization and Equipment for Army Engineer Aviation Fire Fighting Platoons (TOE 5-337 and 5-500) specified the Class 135 as standard equipment. The Class 135 was produced in three models, two of which were built on a 4x2 chassis built by Ford and International Harvester, and the third, which is the one we are concerned with, utilizing the 4x4 Chevrolet G-506 chassis. These trucks featured a front-mounted Darley Model H pump that could deliver 100 gallons per minute at 500 psi. Housed within the apparatus bed was a 300-gallon water tank. Slung beneath the front corners of the apparatus bed were reels for booster lines. A third reel was centered on the rear of the truck, recessed beneath the hose bed. Each reel held 100-feet of 1-inch booster line.

Apparatus bodies were built by American LaFrance, Mack, W. S. Darley, Oren Fire Apparatus, and Central Fire Trucks Corp.

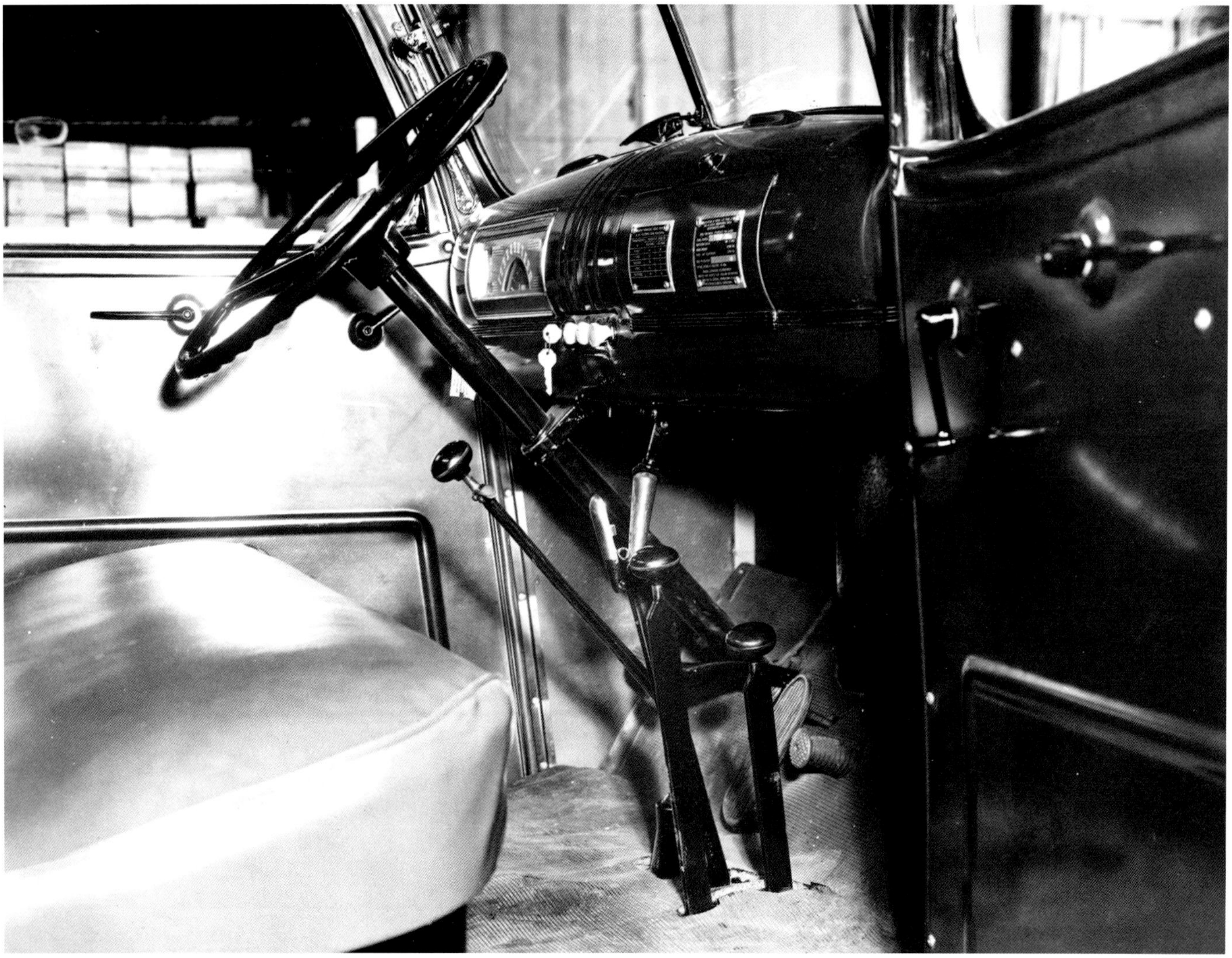

The interior of the cab of a Class 110 Airfield Crash Truck, likely registration number W-50702, is seen through the right door. It has a typical early-G-506 truck-cab layout, with civilian-type instrument cluster and dashboard. General Motors LLC

Four 50-pound CO_2 cylinders were mounted transversely in the lower forward part of the crash body of the Class 110. The valve ends of the cylinders pointed to the left side of the body and were accessed through the two doors toward the front of the body. The hoses from the cylinders were routed to the 100-foot, 1/2-inch high-pressure hose stored on the reel at the rear of the body. Below the rear door is the pump-flushing connection. General Motors LLC

The registration numbers assigned to these 716 trucks were:

W.S. Darley	50	506089-138
Mack Manufacturing Corp	100	506239-338
Oren Fire Apparatus	31	507347-377
Minnesota Fire Equip Company	50	507478-527
Central Fire Trucks Corp	50	507528-577
Minnesota Fire Equip Company	25	507578-602
Mack Manufacturing Corp	150	507843-992
Oren Fire Apparatus	100	508332-431
American LaFrance Foamite	60	5010029-088
W.S. Darley	100	5010139-238

W.S. Darley of Chicago was issued contract 2337-ENG-1098 in February 1943 and the contract was completed in July of that year.

The North Atlantic Division of the Engineers issued Mack contract number W-30-082-ENG (MSP-827) calling for the production of 150 Class 135 trucks. These vehicles were produced from March 1943 through 25 August 1944. Production of 100 additional Class 135 trucks by Mack was accomplished in the final days of December 1943. These trucks, along with 100 examples of Class 125 trucks, were on contract W-2337-ENG-1099.

American LaFrance was issued contract 2337-ENG-1159 in March 1943, and the contract was completed in August of that year. Minnesota Fire Equipment was awarded contract 23065-ENG-31 for Crash Fire Trucks in July 1943, and the contract was completed in December of the same year. In September 1943 St. Louis-based Central Fire Trucks received contract 23065-ENG-121 for

Above: *Another Class 110 Airfield Crash Truck, registration number W-50729, is on display. A spotlight with a chrome housing was mounted on each of the handrails above the crash body, and another spotlight was placed atop the cab. Note the valve handle below the bottom of the crash body to the front of the rear fender.*

Left: *Alongside Class 110 Airfield Crash Truck registration number 501381, a firefighter in an asbestos suit is holding the nozzle attached to the 1/2-inch high-pressure hose, preparing to dispense CO_2. This particular vehicle was equipped with a LaFrance pump instead of the Hale pump used in earlier Class 110s.* University of North Texas Libraries

The Class 135 Airfield Crash Truck reportedly was the most commonly used crash truck at operational airfields in World War II. One version was based on the G-506 Chevrolet cab and chassis, while other versions were based on International and Ford 4x2 trucks. The Class 135 included a semi-high-pressure pump on the front of the vehicle, and containers for 300 gallons of water and 20 gallons of foam. Under each side of the crash body to the rear of the cab was a reel with 150 feet of 3/4-inch high-pressure hose. The vehicle shown here is registration number 506239. Ted Heinbuch collection

crash trucks, and the work on that contract was finished in February 1944. So far no details of the Roanoke, Va-based Oren Fire Apparatus contract have surfaced.

Class 300

The Class 300 truck were equipped to fight brush or forest fires, a function for which the 4x4 chassis was ideally suited. The trucks were assembled by the Central Fire Truck Company, and were equipped with W.S. Darley Model F-300 front-mounted pumps. Because of the type of service that these trucks were intended for, they did not carry any ladders, but they did carry a 10 gallon drinking water container, four 5-gallon back pack water extinguishers, 7 shovels, ten fire buckets, 5 brush axes and five forest fire beaters, and a myriad of other tools for fighting non-structural fires.

The apparatus bed enclosed a 250-gallon water tank, and could accommodate 1,000 feet of 1 ½-hose in the hose bed. On top of the bed were a pair of hose reels, each of which held 150-feet of 1-inch booster line.

Class 325

While the Class 135 was optimized as a "Crash Fire Truck" with a relatively high pressure, yet low volume pump, the visually similar Class 325 was optimized as a general purpose fire pumper. TM 9-2800 lists the vehicles as "Truck, Fire, Powered, Pumper, Class 325, 4x4, 300 GPM." The Barton F-300 pump on these trucks could deliver either 300 gallons per minute at 120 psi, or 150 gallons per minute at 200 psi. Some of the trucks utilized a Darley pump with comparable capabilities.

Unlike the similar Class 135, the Class 325 was not equipped with a rear-mounted booster reel. The trucks did however have two booster reels mounted under the front corners of the bed, much like the Class 135. Each reel held 150-feet of one-inch hose. The bed housed a 300-gallon water tank.

The Class 325 trucks were assembled by W.S. Darley; General Detroit Corp., American Fire Apparatus; Hahn; Oren; and Maxim Motors of Middleboro, Mass.

Class 525

The final development in the conventional cab Chevrolet 4x4 fire truck series was the Class 525 truck. These trucks were an effort to inexpensively field a vehicle comparable to the Class 500 4x2 trucks. These trucks, assembled by Oren, utilized the W. S. Darley F-500 centrifugal pump, and were optimized for use against structure fires. The front-mounted pump, driven off the engine crankshaft, could operate at 250 gallons per minute at 200 psi, or it could deliver 500 gallons per minute at 120 psi.

While like the Class 135 and Class 325, the Class 525 had front bumper mounted pump, the configuration of the pump mounting was different than that of the earlier trucks, thus the front bumper was remounted to a high than normal position in order to provide clearance for the hard suction line. The single booster reel was mounted at the center rear of the truck bed, and on it was 150-feet of booster line.

The body included a 150-gallon booster tank, as well as provision to store 1,000-feet of 2½-inch hose.

G-7153

In addition to the Army, the Navy also obtained fire trucks based on the 1 ½-ton 4x4 Chevrolet. The 103 trucks produced for the Navy, however, were unique in that they were a cab over engine design with a 175-inch wheelbase. Because they were Navy trucks, they were did not carry an Army Class number. These trucks were assembled by Howe Fire Apparatus of Anderson, Indiana, and were built on Navy Contract 95703. The trucks were equipped with a 300-gallon water tank. These 1942-model trucks carried the Chevrolet serial prefix of NX, and were Chevrolet NX model trucks. The chassis were built at Norwood and Baltimore between January and October 1942.

Above: *Class 135 Airfield Crash Truck registration number 506239 is displayed. Two spotlights were mounted on the cab roof, and two were on the handrail across the rear of the crash body. A carbon tetrachloride fire extinguisher, a CO_2 fire extinguisher, a 16-foot combination ladder, and a pike pole are stored on the side of the crash body. A siren, service headlight, and blackout headlight are on the left fender.* Ted Heinbuch collection

Left: *Recessed into the rear of the Class 135 truck was another reel with high-pressure hose. A spacious running board with diamond-tread surface was mounted to the rear of the truck. There was a vertical handrail on each side of the rear of the body, and a horizontal handrail above the rear of the body.*

Above: *Class 135 Airfield Crash Truck registration number 507361 is shown. A noticeable change, a vertical grab handle not seen in the preceding photos of Class 135 vehicles, is mounted on the cab to the rear of the upper part of the door. A storage compartment is to the rear of the dual rear wheels.* Ted Heinbuch collection

Below: *A Class 135 Airfield Crash Truck is viewed from the right side. By now, the cab-mounted spotlights were being mounted even with the upper front corner of each door. A fireman's axe is stored on holders on the side of the body to the rear of the two fire extinguishers. Two fire extinguishers are on holders on the fender. The complement of fire extinguishers for the Class 135 was two 15-pound CO_2, one one-gallon carbon tetrachloride (CTC), and one four-gallon pump-tank.* Ted Heinbuch collection

Above: *The semi-high-pressure pump on the front of the Class 135 Airfield Crash Truck is the focus of this photograph. These pumps had a capacity of 60 gallons per minute at 350 pounds pump pressure. This particular pump is the Darley H100. A service headlight and a blackout marker lamp are on each front fender; in addition, a blackout headlight is on the left fender, between the siren and the service headlight. Ted Heinbuch collection*

Below: *On the right side of this Class 135 Airfield Crash Truck body are, from front to rear, a fire axe, first-aid kit, carbon tetrachloride fire extinguisher,* CO_2 *fire extinguisher, and two combination fog and straight-stream nozzles for fire hoses. A pike pole is stored on yoke-type holders above the crash body. Note the plumbing and the valve handle below the fuel filler to the rear of the cab. Ted Heinbuch collection*

Above: *The rear end of a Class 135 Airfield Crash Truck is shown. A high-pressure hose is on the reel in the recessed compartment at the center. On each side of the running board is a 15-pound CO_2 fire extinguisher. Stored vertically, inboard of each fire extinguisher, is a foam applicator nozzle. Below the platform of the crash body is an opening for accessing the interior of the body. Ted Heinbuch collection*

Below: *The Class 300 Fire Truck was issued to Army bases where there was a possibility of structural fires, brushfires, or forest fires. The G-506 cab and body vehicles were used as the basis for many of these, as were Ford and Chevrolet 4x2 cab and body vehicles. A Darley F-300 centrifugal pump with a capacity of 300 gallons per minute at 120 psi or 150 gallons per minute at 200 psi was mounted on the front end, and a 200-gallon water container was housed in the body. Two reels for 150 feet of 1-inch fire hose were mounted atop the body, and there was a capacity for 1,000 feet of 1 1/2-inch fire hose in the body. Richard Adelman collection*

Above: *The Class 300 Fire Truck had running boards all along the sides and rear of the body. A long rack was on the right side of the body above the rear fender. A large spotlight was mounted above the front of the body, and two smaller spotlights were on top of the stanchions for the handrail at the rear of the body.* Richard Adelman collection

Below: *Like the Class 300 Fire Truck, the Class 325 "Oversea" Fire Truck was intended for use in combating structural fires or wildfires at forward bases with sparse or no water availability. It is said to have been the most common fire truck in U.S. theaters of operations. Features included a front-mounted, centrifugal 300-gallon-per-minute pump and a 300-gallon water tank. The G-506 Chevrolet cab and body was one of several vehicles on which the Class 300 bodies were installed, and was the only 4x4 model.* US Army Engineer School History Office

Above: *A reel containing 150 feet of 1-inch high-pressure hose was under each side of the Class 325 Fire Truck body, to the front of the rear wheels. A tray was below each reel. A holder for a fire extinguisher is on the rear end of the rear fender. Note the brace for the rear running board.* US Army Engineer School History Office

Below: *On the front of this Class 325 Fire Truck, probably made by General Detroit, is a Darley centrifugal pump: apparently the F-300 model, with two hose connections on the top. There is a loop-shaped extension on the left side of the brush guard, to protect the blackout headlight. Note the elongated opening in the front of the bumper to allow access to the suction coupling at the bottom front of the pump.* US Army Engineer School History Office

Above: *Unlike the airfield crash trucks discussed earlier, the Class 325 Fire Truck lacked a hose reel on the rear of the body. Similar to the Class 300 Fire Truck, the Class 325 had a platform on top of the body, with an opening at the rear of the body under the platform.* US Army Engineer School History Office

Below: *Stowed along the right side of the apparatus bed of the Class 325 were two ten-foot sections of 3-inch hard suction line laying adjacent to each other, and a 20-foot unpainted wooden extension ladder, with a 12-foot roof ladder stowed just inboard.* Ted Heinbuch collection

Mounted on the front of this Maxim-assembled Class 325 is a Darley F-300 fire pump, one of the standard pumps fitted to Class 325 trucks. The pump was driven off the front of the engine crankshaft via a short driveshaft. Because the pump was coupled to the front of the crankshaft, provision was made for the engine hand crank to engage the fire pump, rather than the end of the crankshaft. Ted Heinbuch collection

Above: *Stowed on the driver's side of the apparatus bed were an array of man-portable firefighting devices ranging from simple galvanized buckets to 4 and 5-gallon pump and back-pack fire extinguishers.* Ted Heinbuch collection

Below: *Secured to the rear step of the Class 325 were a variety of adapters, couplings, nozzles and fireplug wrenches.* Ted Heinbuch collection

Above: *Although no longer in US Army service, this Class 525 was still in Federal service, albeit with the Department of Interior, when photographed. The unique, elevated mounting of the front bumper characteristic of this class is plainly visible.* Richard Adelman collection

Below: *Unlike the Class 135 and Class 325, which had booster reels under slung beneath the front corners of the bed, the Class 525 had stowage compartments there, as well as beneath the rear corners of the bed.* Richard Adelman collection

Above: *In 1942 the U.S. Navy obtained 103 examples of a version of fire truck based on the Chevrolet Model G-7153 cab-over-engine 4x4 truck. Since this was a Navy truck. Since this was a navy truck, it did not carry a Corps of Engineers class designation. The body was produced by the Howe Fire Apparatus Company of Anderson, Indiana.* National Archives

Below: *The vehicle had a large spotlight on the cab roof, several hose reels on top of the body, storage for 300 gallons of water in the body, and four fire extinguishers on the side of the body. A spare tire was stored on the side of the body at the rear.* National Archives

Above: *Vintage adverts.* General Motors LLC

Chapter 10

Panel Trucks

The first production of the panel truck model of the 1½-ton Chevrolet was given Master Book Symbol YG, although it carried the serial prefix of YP. The Master Book Symbol assigned to the second production trucks was ZG, and when the 1942 model trucks were introduced, the Master Book Symbol NG was used.

The first of the panel delivery trucks, the YG vehicles, were bought on contract W-398-QM-8278, and though not among the vehicles called for in the initial specifications, were delivered during the second month of series production. Twenty-two examples were accepted by the Army in December 1942. Through the course of production, the panel trucks were built in Norwood, Tarrytown, Baltimore, Oakland, and St. Louis.

These panel van trucks are most widely known for their use as by the Signal Corps, which assigned them the designation K-51, and used them to house the SCR-299 radio set. To power the radio equipment, the vehicle towed a K-52, which was a 1-ton trailer outfitted with a PE-95 generator.

Signal Corps efforts to procure a radio system with the capabilities that would be found in the SCR-299 had begun July 1940. Following the May 1940 Louisiana Maneuvers (an exercise overshadowed in history annals by the 1941 maneuvers of the same name), Colonel Dawson Olmstead, who later became the Chief Signal Officer, advocated for the rapid development of a truck-mounted high-powered radio system. So passionate was Olmstead about this that his colleagues on the Signal Corps Board, over which he presided, dubbed the project "Olmstead's baby."

Chevrolet 1 1/2-ton 4x4 Panel Delivery Truck registration number W-322446, is a very early production vehicle with the battery beneath the cab floor. The small feature on the body to the rear of the cab door handle was a forward-pointing reflector, a feature not seen in later production panel delivery trucks. The dark object directly below the running board was a door sill on a building in the background, most of which was airbrushed out. National Archives

Above: *A registration number is not visible on the hood of this G-506 panel delivery truck, but the Chevrolet maker's mark stamped on the side panel of the hood is faintly visible, dating this vehicle to earlier than mid-1943. This is a later production vehicle and has an cutout in the running board apron to provide access to the fuel filter.* General Motors LLC

Below: *Like the other G-506 trucks, the panel delivery truck had dual rear tires. A spare tire was stored on the left side of the body. This specimen, registration number W-3115328, was manufactured under contract number W-3989-QM-12396. A taillight with two exposed electrical cords was on each side of the rear part of the body. Note the graceful lines of the recessed panel on the upper half of the side of the body, which also extended around the window in the cab door.* General Motors LLC

Above: *A Chevrolet 1 1/2-ton 4x4 Panel Delivery Truck with no registration number on the hood is viewed from the left side in a photograph dated 1941. The small reflector to the rear of the cab door casts a shadow on the truck body. Note that by now there also was a similar reflector on the rear of the body, below the taillight; this reflector pointed to the rear, and its shadow is visible.* Military History Institute

Below: *Chevrolet built a panel delivery truck version of the G-506. This was designated in succession the G-4112, G-4105 and G-7105 and featured a fully enclosed body with a continuous roof from the cab to the rear. The body was based on the one used on Chevrolet's line of 1/2-ton and 3/4-ton 4x4 panel trucks. The battery was set in a recess at the bottom of the body, and the fuel filler was to the rear of the battery. This example, registration number W-337303, was built under contract W-398-QM-9402 and was accepted by the Army in 1941.* National Archives

Above: *Markings for the 161st Signal Photographic Company are on the cab door and the bumper of Chevrolet 1 1/2-ton 4x4 Panel Delivery Truck registration number W-356761. On the left side of the bumper is painted "No. 41," a reference to the truck's position in its unit's order of march. The Chevrolet maker's stamp is slightly visible on the side panel of the hood, and to the rear of the spare tire is a vertically mounted pioneer tool rack.* National Archives

Below: *In a view through the open rear doors into a Chevrolet 1 1/2-ton 4x4 Panel Delivery Truck, the internal framing of the body is visible, along with the two seats in the cab and the rear wheel wells. The floor was constructed of oak planks running fore and aft. Below the sill of the rear doors were two bumperettes with a tow pintle between them. A doorstop is at the center of each door.* National Archives

That radio system, which was made by the Hallicrafters Company of Chicago, was designated the SCR-299. Hallicrafters, which was one of the largest manufacturers of amateur radio equipment in the world at the time, produced a powerful system, the HT-4, which had captured the attention of the Signal Corps.

While the Signal Corps worked to develop their own 100-mile range mobile radio set for use by the Armored Force, the AF-I (based on the SCR-197, this was first designated SCR-505, and later SCR-597), modification and adaptation of the HT-4 seemed to be potential stop-gap system. That adaptation, the SCR-299, was a mobile High Frequency station capable of Continuous Wave (telegraph) or voice communication (AM). The heart of the system was the BC-610 transmitter, a militarized version of the civilian Model HT-4. The transmitter produced 400 watts output CW telegraph and 300 watts voice, operated on 2-20 MHz, and had a range of about 100 miles. To adapt the HT-4 to military service, the steel cabinets were strengthened and shock proofed, clamps added to secure the many tubes in place, guide slots added to aid in the rapid replacement of components, and corrosion-resistant coatings added to some internal components. Various relays were added to allow automatic changeover of military-operations circuits. A chatter-proof relay system was designed to prevent relay kick out during rough terrain operation. Various reset controls were relocated to the front panel to ease operation.

A master oscillator was added to provide continuous coverage without a maze of crystals, each ground to the one of many frequencies required for military communications. A control on the tuning unit allowed the switch from crystal to master oscillator. A 35-foot vertical whip antenna was provided, and a special antenna coupler used to establish accurate loading at any of the transmitter's wide frequency ranges.

However, transmission is only half of what the system was intended to do – reception was the other component of its function. To accomplish this, one BC-312 and one BC-342 military receiver were each installed, along with a BC-614 speech amplifier, BC-729 (tuning unit), BC-211 (frequency meter), and RA-63 rectifier. Many spare parts and tools for both the receivers and

Dozens of completed Chevrolet 1 1/2-ton 4x4 Panel Delivery Trucks are lined up on the dirt floor of the International Amphitheatre, home of the International Livestock Exposition, in Chicago during World War II. These trucks, along with thousands of crates of supplies and other war materiel, were being stored at the Amphitheatre, since this venue had ceased hosting its usual fare of indoor races and livestock shows for the duration of the war. General Motors LLC

Above: *The Chevrolet K-51 was a U.S. Army Signal Corps radio truck based on the 1 1/2-ton 4x4 Panel Delivery Truck. It carried an SCR299 radio set and was designed to be used in conjunction with the K-52 Trailer, which carried a PE95 Power Unit for the radio equipment. A distinctive feature of the K-51s was the presence of a duffle cabinet on each side of the body, for stowing crew gear, supplies, and equipment. This K-51, registration number 365565, is fitted with a tied-back transmitting antenna, two freestanding receiving antennas, and a wire reel.* General Motors LLC

Below: *On this K-51, registration number 356858, which the Army accepted in 1942, the hinges on the door on the left duffle cabinet are reversed compared to the door seen in the preceding photo. Here, the hinges are on the rear edge of the door, and the padlock hasp is on the forward edge. In addition to the padlock hasp, on the front edge of the door are two cam locks.* Communications-Electronics Command History Office

Above: *Chevrolet K-51 registration number 3115096 is viewed from the rear. Painted on the cab door and faintly visible is "K-51." A reel of communications wire is stored on each side of the body. Note the retainer chain with a turnbuckle at the top of the right-hand reel; this chain was in addition to the spindle the cable reel was stored on. Eyebolts toward the rear of the roof were supplied for the tie-down cord for the transmitting antenna.* General Motors LLC

Below: *Chevrolet K-51 registration number 337315, which was accepted in 1941 and was produced under contract number W-398-QM-9402, is rigged for travel, with a K-52 Trailer hitched behind it. Visible on the roof of the K-51, above the duffle cabinet, is a ventilator. On this vehicle, a small storage box is to the immediate front of the duffle cabinet.* Communications-Electronics Command History Office

transmitter were stored in the CH-88 chest and CH-89 bench within the shelter. The entire radio system could be operated remotely from EE-8 field phones from up to a mile away.

As mentioned, the radio system was connected to either an external power supply, or more often the trailer-mounted PE-95 generator. Multiple versions of the PE-95 were built, notably by Onan, with the –A, –B and –C models powered by a Ford engine, and the –E, –F and –G being powered by Willys engines. A tether

Above: *This composite photograph shows key elements of the interiors of the Chevrolet K-51 Panel Truck and the K-52 trailer. The SCR-299 Radio Set took up most of the interior of the truck to the rear of the cab, with room for the operators' bench seat (not shown here) and other equipment. Inside the K-52 Trailer is the PE-95 Power Unit, which supplied electrical current for the radio equipment.* General Motors LLC

Below: *A Signal Corps photograph dated 15 October 1941 shows the interior of a K-51 Panel Truck with an SCR-299-T2 Radio Set installed. The quarters were cramped, and there was not much headroom for the operators. Below the operator's table are stowed two EE-8 portable field telephones, and just inside the rear door at right is a fire extinguisher.* Communications-Electronics Command History Office

Above: *The interior of a K-51 Panel Truck is viewed through the rear doors at a base at Bristol, England on 27 January 1943. The photo is credited to the 56th Signal Battalion, to which presumably this vehicle was assigned. This photo provides a rare view of the inboard face of the storage cabinets mounted on the right side of the compartment. Part of the right seat in the cab is visible in the background.* National Archives

Below: *The various components of the SCR-299-A, -B, and –C as installed in the K-51 Panel Truck are displayed. BC-312 and BC-342 Radio Receivers as well as a BC-731 Control Box and other accessories were on the table, while the large BC-610 Radio Transmitter, with the antenna tuning unit mounted on top of it, was mounted just behind the cab seats.*

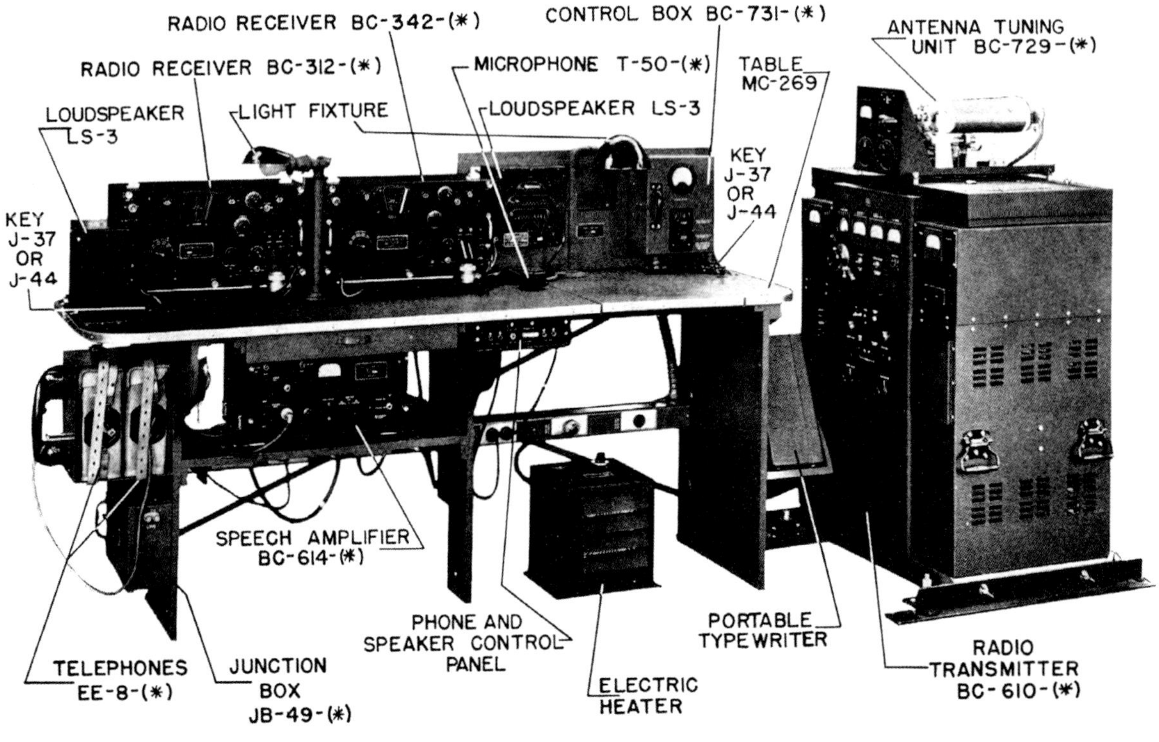

ABOVE: *“The SCR-299 Radio Set in a Chevrolet K-51 Panel Truck was photographed at Camp Gordon Johnston, Florida, on 23 February 1943. Also present are the operators’ bench seat, which doubled as a storage box, and a fire extinguisher. Note the curved cross members of the ceiling frame.* National Archives

BELOW: *Some K-51 Panel Trucks had the SCR-299-D Radio Set, as depicted here. The layout of components on the table was different than with the SCR-299-A to –C models; for example, the speech amplifier was moved up to the table and the two radio receivers were mounted on each side of that amplifier.*

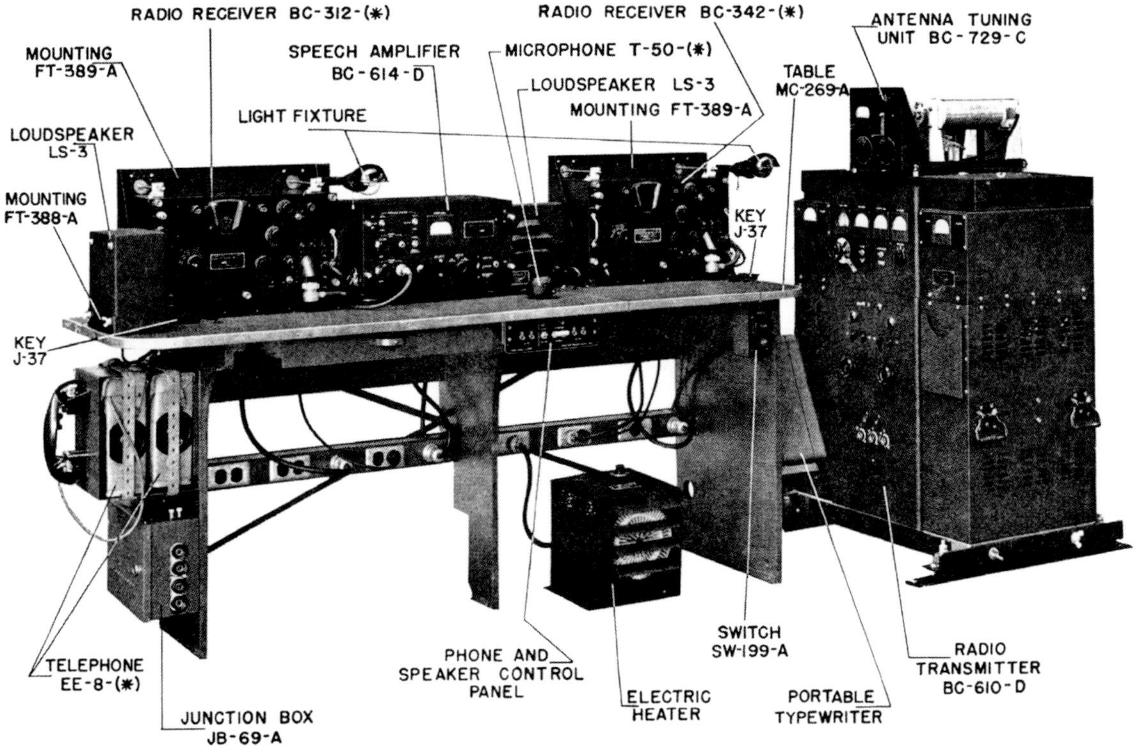

A civilian technician is working on the left receiving antenna base on a K-51 Panel Truck. Next to him on the roof of the vehicle is the ventilator hood: an important piece of equipment, as the SCR-299 Radio Set built up a lot of heat during operation. Another K-51 is parked to the right side of this vehicle. Jim Gilmore collection

connected the K-52 generator to the K-51 truck, allowing the generator to be operated from within the truck, and the entire system transmit and receive while in motion. The generators were notably reliable, and very much in demand not only for powering the SCR-299 as intended, but other uses as well.

Development of the SCR-299 took only 18 months, and as testing neared completion, the 1941-dated Fifth Supplemental Program for Fiscal Year 1942 included funding to purchase a number of the sets, at $15,000 each.

While the original purpose of the SCR-299 was to satisfy the requirements of the Armored Force, in the words of UNITED STATES ARMY IN WORLD WAR II, The Technical Services, The Signal Corps, The Test, "The 299 came to be used by everyone and his brother, not just by the Armored Force alone, but also by the Infantry, by the Air Forces, and by the Allies. Production began in the early spring of 1942, when Hallicrafters delivered the first model in March and more than a hundred in April."

"The SCR-299 could travel anywhere its truck transport and trailer, carrying its excellent power plant, PE-95, could negotiate. Its mobility, then, accorded perfectly with the needs of America's new motorized armies. Moreover, its power and reach were impressive, not just "100 Miles in Motion" for which the Signal Corps Laboratories had developed it. On selected high

Standing on the left fender of the K-51, the technician makes further adjustments to the left receiving antenna. The lead-in wire from the antenna through a grommet in the body of the truck also is visible. With the reel of communications wire removed from the side of the body, the mountings for the reel are in view; this is also the case with the other K-51 in view. Jim Gilmore collection

frequencies using sky waves, the 299 could fling Morse code signals for many hundreds of miles; this, of course, in addition to serving as a reliable radiotelephone up to 100 miles."

While it had been planned that four SCR-299s be deployed with the 835th Signal Service Company, to the China-Burma-India theater when the unit sailed on 19 March 1942, the unit left without its planned SCR-299s, as the production units were not available for issue until 31 March. However, several SCR-299s were used in Operation Torch, with six in Oran, four in Casablanca and two in Algiers. One of the SCR-299s in the Oran area by Company B of the 829th Signal Service Battalion established and maintained a direct communications channel with England – a distance of 2,800 miles to the north, impressive performance for a radio system engineered for a range of 100 miles. The same unit, with the same equipment, also maintained contact with Accra, Ghana, 4,800 miles to the south. This was done purely with the issued gear.

In a memo to the Chief Signal Officer, Major Lawrence C. Sheetz wrote, "The SCR-299 is far and away the best field radio equipment in the theater," further stating "this is the unanimous opinion of all British and American officers.... "One British staff officer stated that "using a 299 after using any British R/T set is like driving a private car after handling a three-ton lorry."

A K-51 was among the nearly 1,000 items displayed at the Army Service Forces Exhibit at the San Antonio Depot as part of the Fourth War Loan Drive in February 1944. Standing on and near the K-51 were, left to right, O.L. Butler, T.C. Frost, Creston Funk and H.B. Housman, all heavily involved in the War Loan Program. The K-51 has been fitted with a communications-wire reel painted in a very dark color.
Jim Gilmore collection

While Hallicrafters was based in Chicago, the radio systems for the SCR-299 were produced in the government-owned plant in the Chicago suburb of Clearing, Illinois. This 200,000 square-foot building had been built in the late 1920s to handle Buick Motor Company's parts and vehicle distribution for the greater Chicago area. Located at 5025 West 65th St at South Lavergne Ave., the half-block facility was served by five railroad sidings. During WWII, this facility, operated under the auspices of the Chicago Signal Depot, assembled assembled the SCR-299s and installed them in the trucks and trailers. By early 1943 the depot had shipped 1,300 completed sets to tactical units in this country and overseas, as well as 125 each to the British and to the Russians, and three each to China and Mexico. All those for Lend-Lease were boxed in waterproof boxes weighing 16,800 pounds. Outside the building was 180,000 square feet of open storage space, where 1,000 more trucks waited for SCR-299 installations.

For some time researchers and enthusiasts have been baffled by another 1½-ton 4x4 Chevrolet panel truck, the K70, about which scant information is available. The February 1943 War Department Circular number 180, Responsibility for Procurement, lists vehicle and body type, as well as indicating what agency has responsibility for the each of the three categories; Chassis, Body and Storage and Issue lists the K70 as "Development-Should not be listed." Since production of the panel truck as a whole ended in December 1943, it is probable that this truck never entered production.

Above: *Microphone in hand, an operator is at work at the SCR-299-C Radio Set inside a K-51 Panel Truck. Below the four-inch-thick seat cushion is an equipment chest, and more equipment is stored on the right side of the vehicle, behind the radio operator. Note the stencil on the right door identifying the type of radio equipment in the vehicle and the radio manufacturer, Hallicrafters Co.* General Motors LLC

Below: *A Signal Corpsman is about to install a SCR-299 transmitting antenna into its base atop a K-51 Panel Truck. The antenna was assembled in sections, with the bright-colored couplings between the sections being somewhat visible in this photo. The two receiving antennas, visible here, also were assembled in segments.* Jim Gilmore collection

Above: *In a view looking over the top of a PE-95 Power Unit in a K-52 Trailer, a soldier is handing a dispatch to the operator of an SCR-299 Radio Set in a Chevrolet K-51. As was typical, information on the specific type of SCR-299 radio equipment carried in the panel truck is stenciled in black on the inside of the right rear door.* Jim Gilmore collection

Below: *A radio operator, the same Signal Corpsman seen in the preceding photograph, is at work at the radio inside a Chevrolet K-51 Panel Truck. This set is the SCR-299-D, as indicated by the presence of the BC-614-D Speech Amplifier to the right of the BC-312 Radio Receiver, which is on the far left of the desk. In previous models of the SCR-299, the speech amplifier had been below the desk.* Jim Gilmore collection

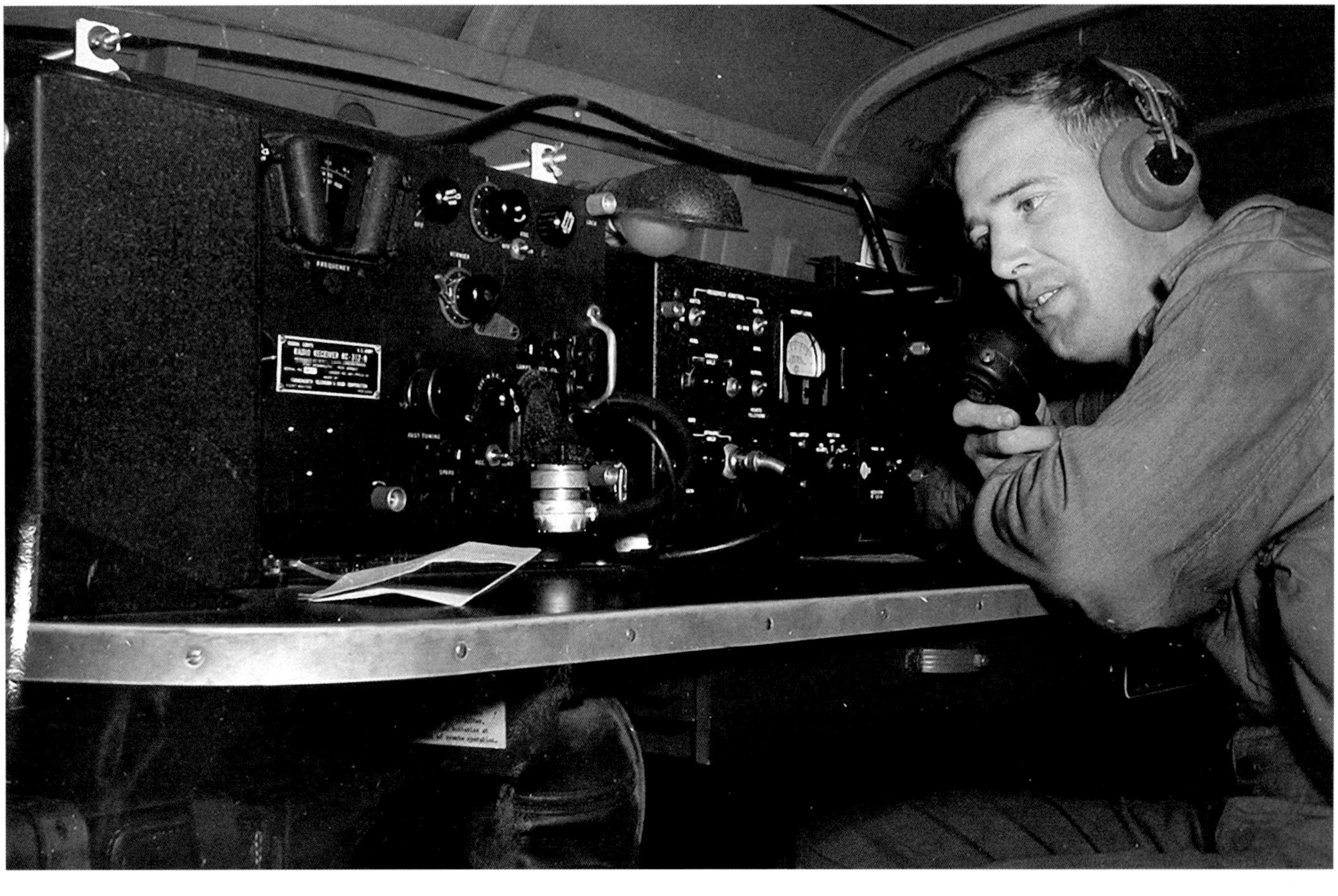

Above: *Another Signal Corps radio operator is conducting voice communications over an SCR-299-D Radio Set in a K-51 Panel Truck. This soldier and the one in the preceding photo are wearing a bag containing a gas mask, with the insignia of the Chemical Warfare Service stenciled on it.* Jim Gilmore collection

Below: *This well-weathered Chevrolet K-51 Panel Truck is parked with a K-52 Trailer hitched to it is parked in the Ford-operated Chester Tank Depot. The truck bears registration number 3115275 on the hood, along with a letter "S" indicating the vehicle was equipped with radio-interference suppression equipment. A good view is available of the distinctive roof vent of the K-51.* National Archives

Above: *A PE-95 Power Unit, for generating electrical current for an SCR-299 Radio Set as mounted in a K-51 Panel Truck, is seen in a K-52 Trailer, which carried a PE-95 Power Unit for the radio equipment. Below the grille is the muffler. The photo was taken at the Chester Tank Depot in Chester, Pennsylvania, around January 1, 1943.* National Archives

Below: *Several dozen G-506 Chevrolet Panel Trucks are in view in a warehouse at an unidentified location. Several of the trucks on the right side have the distinctive roof vent of the K-51, and at least the first vehicle on the left is a K-51; note the duffle cabinet on the side. Registration numbers are present on the hoods but none are fully legible: several are in the 310XXXX numerical range.* General Motors LLC

Above: *The U.S. Army Signal Corps developed this repair truck, based on the Chevrolet 1 1/2-ton 4x4 Panel Delivery Truck. As photographed in New York City on 4 May 1942, the vehicle had a support structure extending to the left rear of the chassis frame, to support a davit with a chain hoist. The chain hoist has a cover over it in this photo.* National Archives

Below: *The Signal Corps repair truck is viewed from the left rear, showing the davit positioned to hoist a load. The tapered boom at the top of the davit consisted of two pieces of steel with a space between them; they were welded to the top of the vertical pedestal of the boom. A close examination of this series of photos shows that the pedestal was attached to the vehicle solely by the structural-steel frame mounted below the rear of the truck body.* National Archives

Above: *The davit and the Yale Spur-Geared Block chain hoist of a Signal Corps repair truck is being employed in a demonstration of the hoisting of a Kohler generator at New York on 4 May 1942. Visible inside the vehicle is an upholstered, rear-facing seat and a workbench with drawers; atop the bench is attached a vise.* National Archives

Below: *The repair truck is viewed from a slightly different perspective with the Kohler generator suspended above the floor. In this photo, the seat inside the compartment has been removed. A clear view is available of the diagonal brace for the bottom of the davit.* National Archives

Above: *The Kohler generator has been loaded into the rear of the repair truck with the assistance of the davit. Inside the vehicle are shelves and parts bins to the left, and a work bench and racks with electronic equipment to the right.* National Archives

Below: *The interior of the rear of the Signal Corps repair truck is seen without the generator loaded. The cabinetwork inside this truck was of unpainted wood, and the workmanship and finish was somewhat crude, with some rough-cut lumber being in evidence. The small panel on the side of the workbench was a hinged base for attaching a vise.* National Archives

Above: *The workbench and on the right side of the repair truck included six drawers in two groups, three high. The bench was mounted to the floor with L-shaped brackets. Mounted atop the bench was a rack holding what appears to be electronic test gear. Electrical receptacles and conduits are on the top and the side of the rack; a receptacle also is visible near the lower center of the rack.* National Archives

Below: *In a photo taken on 4 May 1942 in New York City, a man is demonstrating the working conditions within the Signal Corps repair truck. There was little or no headroom for a person standing up inside the compartment. A vise has been mounted on the hinged panel on the side of the workbench, supported by a brace. Note the two angle irons attached to the rear of the floor. These were not present in the photos of this truck taken on 30 April 1942 and evidently were installed as retainers for the base of the Kohler generator, to keep it from shifting around during travel.* National Archives

Above: *As seen in a 30 April 1942 photo, a ladder was on a rack on the right side of the Signal Corps repair truck, and there was a storage box along the lower part of the body, which partially covered the battery box.* National Archives

Below: *A mechanic is straddling a loading dock and the rear of the platform of a Signal Corps repair truck at New York City on 8 May 1942. It is not clear if this repair truck was a one-off experimental or prototype conversion, or if the Signal Corps produced numbers of these vehicles.* National Archives

Chapter 11

Airfield Lighting

Type J-3 Field Lighting Truck

One of the first, if not the first, of the specialized trucks built on the 1½-ton 4x4 Chevrolet chassis were the J-3 field lighting trucks. Specification USA-LP-91-701J outlined the characteristics for the chassis for these trucks, while the body was covered by specification 30138. The platform bodies for the trucks were built were built by Chevrolet, and the American Gas Accumulator Company of Elizabeth, New Jersey installed the lighting equipment. These trucks were specifically intended to light temporary and training airfields.

Air Corps contract W-535-AC-16747, issued in November 1940 was the first to call for production of these vehicles. The contract was completed by the end of March in the following year. This was followed by June 1941 contract W-535-AC-19340, which was completed by December 1941. The 142 trucks on this contract used "NF" as both the serial prefix and Master Book Symbol, as did the 100 similar trucks built for Defense Aid on contract DA-W-535-AC-888. While the Defense Aid trucks did not have US Army registration numbers, the trucks on W-535-AC-19340 bore registration numbers W-609058 through W-609199.

To meet the U. S. Army Air Forces' needs for mobile lighting to temporarily illuminate airfields for nighttime flight operations, the Chevrolet G-7143 Field Lighting Truck was developed. Based on the Chevrolet 1 ½-ton 4x4 chassis and cab with 145-inch wheelbase, the G-7143 came in three versions: the virtually identical Types J-3 and J-4 (they varied only in the power plants), featuring six floodlights in box-shaped housings, and the Type J-5, which had four round flood lights. Seen here is a G-7143 Type J-3, registration number W-605505, with a cover over the floodlights. The Le-Roi power plant was mounted transversely between the cab and the lights. National Archives

A Chevrolet Type J-3 Field Lighting Truck is viewed from the upper left, with the lights in the lowered, travel position. The power plant included a generator powered by a gasoline engine. The power plant, enclosed in a sheet-metal housing, was mounted transversely across the body. The floodlight assembly was designated the American Gas Accumulator Co. (AGA) Type PMF 9-6 Mobile Floodlight. General Motors LLC

The Navy issued contract NOs-95055 for $887,000 in December 1941. This contract covered the production of 71 further NF chassis for J-3 airfield lighting trucks. Only six of these were to be for the Army, with the remainder for the Navy. The six Army vehicles were assigned registration numbers 609086, 609092, 609096, 609101, 609104, and 609105. This contract was completed in May 1942.

Type J-4 Field Lighting Truck

The next vehicle in the evolution of the airfield lighting truck was the J-4. These trucks were built in two distinct variations, one based on an International Harvester chassis for the Navy, and another on the Chevrolet chassis, for the Army. The lighting equipment on both types of trucks were identical, and were covered under specification 30150. The lighting equipment was produced by the American Gas Accumulator Company, who also mounted the equipment on the truck chassis. This work was covered under contract W535-AC-31742 for the Army trucks, and NXsa 13629 for the Navy. This $926,000 contract was issued in August 1942, and the work was completed in February 1943. Production of the International trucks for the Navy took place from September 1942 to May 1943.

The business end of the truck were the six floodlights mounted at the rear of the truck, each equipped with a seven-element Fresnel lens. The lamp assembly

As seen from the rear in a photo of a Type J-3 Field Lighting Truck, the J-3 and J-4 versions featured six floodlights on a pedestal which could be raised and rotated to suit the situation. The lamps are shown here in the raised position. A hinged platform is at the rear of the body, in its lowered position. Two storage boxes at the rear of the body are equipped with upholstered seats. The registration number is very faintly visible on the rear cross member of the body" "U.S.A." to the left, "W-60" at the center, and "5505" to the right. General Motors LLC

The floodlights of a Type J-3 Field Lighting Truck are viewed in their travel position from the front left. An access door is on the outboard side of each floodlight enclosure. Between the two stacked sets of three floodlights is the panel box, above which are the control wheels. What appears to be a beacon is on the top of each stack of floodlights. National Archives

The floodlights of a Type J-3 Field Lighting Truck are seen from the rear, in their travel position. In the foreground are two cushioned seats, which also served as lids for storage compartments for spare lamps and tools. National Archives

On Chevrolet G-7143 Field Lighting Truck registration number W-606009, the floodlights are in their lowered position. The floodlights on the J-3 and J-4 were produced by the American Gas Accumulator Co., and each J-3 or J-4 unit produced over one million candlepower. Information about the vehicle is stenciled on the driver's door but is not legible. General Motors LLC

was mounted on a pedestal, and the six-lamp light head could be manually raised, lowered and rotated as needed to illuminate the desired area. Each of the six bulbs was a 1,500-watt, 32-volt lamp, with a combined output of the truck equal to over a million candlepower.

Power for these lamps was supplied by an AN2518 10 kilowatt, single-phase generator able to deliver 32, 120 or 240-volts. The generator was powered by a LeRoi D176 4-cylinder engine, which was supplied by its own 20-gallon capacity gasoline tank. The generator itself was made by the Marble Card Electric Company, of Gladstone, Mi.

When packed for overseas shipment, the floodlights as well as the generator were dismounted from the truck and crated.

Type J-5 Truck Field Lighting Unit

The final step on the evolution of the Chevrolet-based airfield lighting truck was the J-5. This truck differs substantially from the previous J-3 and J-4 models. First of all, the lighting equipment was manufactured by the Davey Compressor Company, Kent, Ohio.

The earlier trucks had six-light arrays with Fresnel lens, whereas the J-5 had four 25-inch diameter B-7 lights. Further, the earlier vehicles carried an engine-driven generator to supply current to the lights, but the 10 kilowatt Westinghouse type SK generator of the J-5 was driven by the truck engine via a Davey Model 50-B power takeoff (PTO).

The PTO was installed as an integral part of the transmission to transfer case driveshaft, and was of the multiple V-belt type. A separate V-belt off the PTO drove the governor.

The J-5 airfield lighting trucks were produced under two contracts. The first, W-535-AC-25544, was issued in May 1942 and was completed in September 1942. The second, or contract W-535-AC-32103 was issued in August 1942 and ran through November 1942.

Above: *A Chevrolet G-7143 J-3 or J-4 is viewed from the upper left, with the lights in the lowered position. The power plant included a generator powered by a gasoline engine. The power plant, enclosed in a sheet-metal housing, was mounted transversely across the body.* General Motors LLC

Below: *The J-5 version of the Chevrolet G-7143 Field Lighting Truck featured four floodlights in can-shaped enclosures, on a pedestal mount. These lights, manufactured by the Davey Compressor Co., developed eight million candlepower. Electricity for the lights was from a generator powered by a power takeoff from the truck engine. Unlike the large, transverse self-contained power plant of the Types J-3 and J-4, the bed of the Type J-5 had only a smaller housing for the generator alone, with more room on each side of the housing. This Type J-5 was photographed on September 20, 1943, apparently at Chanute Field, Rantoul, Illinois. Nomenclature and manufacturers' information is stenciled on the cab door.* National Archives

Above: *The U.S. Navy also used the Chevrolet G-7143 Field Lighting Truck to illuminate airfields during nighttime. In this view of a Curtiss SB2C Helldiver at an unidentified airfield, the rear end of a J-3 or J-4, including the floodlights, is visible in the right background. The radiator grille of the power plant is partially in view.* National Museum of Naval Aviation

Below: *A Type J-5 Field Lighting Truck is seen from the rear. The J-5 had a single, wide storage box on the rear of the body, and this truck's registration number, 6061912, is stenciled on it. All of the floodlights have vertically ribbed lenses except the one to the upper left, which has a smooth lens.* National Archives

This G-7143 Type J-5, registration number 606194, has a placard explaining that the vehicle was assigned to the Special Photographic Service Section and was used as supplementary lighting for motion-picture photography. National Museum of the United States Air Force

Crewmen of a Type J-5 Field Lighting Truck, registration number 6060913, are performing maintenance on the floodlights. The hinged lenses on the two upper lights have been opened for inspection, showing the lamps and the reflectors. The platform on the rear of the vehicle facilitated working on the two lower lights. National Archives

Chapter 12

Low-Silhouette Trucks

The profile or silhouette of a vehicle is of concern to military planners for a couple of reasons. One is that the overall height of a vehicle impacts the amount of shipping space required. Secondly and perhaps even more importantly, the taller the vehicle the bigger target it makes, and the easier spotted it is.

Major General R. M. Danford, Chief of Field Artillery, wrote to the Assistant Chief of Staff, G-4 in late May 1941 concerning vehicle demonstrations he had seen a few days earlier at Camp Holabird and Aberdeen Proving Ground. In his letter, Danford lauded the short wheelbase and low silhouette of the ¼-ton 4x4 (jeep) and suggested that those characteristics be applied to vehicles of other weight classes as well. Danford included a proposed list of "Approximate Tentative General Characteristics for 1/2-ton, 1 1/2-ton and 2 1/2-ton 4x4 trucks." Accordingly, the manufacturers were asked to develop low profile 4x4 truck designs. In October 1941 the scope of the project was expanded to include a low silhouette 2 1/2 to 3-ton 6x6 vehicle as well.

Advancing the same notion further, on 1 July 1941 Brigadier General J. E. Barzynski, Quartermaster Corps, wrote to the Adjutant General, advising that the Quartermaster Corps Technical Committee had met 27 June 1941, and at that meeting the committee members had looked over blueprints of new low silhouette, short wheelbase trucks submitted by several vehicle manufacturers, as well as discussing the Ford "Swamp Buggy," the demonstration of which was one of the factors inspiring Major General Danford.

Brigadier General J. E. Barzynski, noted that "It was the consensus of those present that rather than request Ford "Swamp Buggy" vehicles for service test at that time, it would be advisable to delay action approximately 40 days to permit the Quartermaster Corps to secure sample vehicles from other manufacturers for comparison. The following motion was then made, seconded and approved:

"It is recommended that the Quartermaster Corps be authorized to take expeditious action to set up a project for the purpose of development and procurement for service test of improved 4-wheel, 4-wheel drive motor vehicles of 3/4-ton, 1 1/2-ton and 3-ton capacities with the following characteristics: low silhouette, minimum wheelbase, maximum flotation, maximum clearance, minimum weight and cross-country ability. These vehicles should have suitable speed, approximately 55 m.p.h. for the 3/4-ton and 1 1/2-ton capacities and a minimum of 45 mph, with 55 m.p.h. if practicable, for the 3-ton capacity.

The component units of these vehicles should possess such factors of safety as to be free from excessive maintenance. It is further recommended that necessary funds be set up for the procurement of not to exceed fifty (50) of each of the three (3) capacities to determine the practicability of design involved."

In response, on 16 July, J. G. Brackinridge of the Adjutant General's office directed that the Quartermaster General take on a low-profile truck project as described in the Quartermaster Corps Technical Committee motion. Assigned development project number KG-261, the intial report on the task notes; "Preliminary development work was immediately started. Conferences with vehicle manufacturers' representatives were held at the Holabird Ordnance Base (sic). Manufacturers were invited to submit designs and specifications indicative of their own interpretation of the proposed vehicles. Quotations were subsequently requested in conformance with specifications EDT 1-42, for Truck, Improved Development Type, dated 1 March 1942. Bids and drawings were later received from interested manufacturers.

These, together with design prints, were submitted to and considered by the Quartermaster Corps Technical Committee."

As the various manufacturers, including Chevrolet, were working on their designs, the Japanese were planning their attack on Pearl Harbor. The 7 December attack brought about a tremendous shift of priorities within the War Department. Emphasis was placed on increasing production of already-developed models, and the low profile project was not revisited until February 1942. On 19 February 1942 the Purchasing and Contracting Officer, Holabird Quartermaster Base,

Above: *Pursuant to a call by the Chief of Field Artillery for improved vehicles in the latter part of World War II, the Ordnance Research and Development Center undertook Project No. 1103, to develop a fleet of low-silhouette vehicles. These trucks, proffered by various manufacturers in different weight classes, incorporated features that artillery personnel thought were lacking in available production vehicles: chiefly, better concealment and a smaller target for enemy gunners. One of the trucks developed and tested under Project No. 1103 was a low-silhouette prototype vehicle based on the Chevrolet 1 1/2-ton 4x4 Cargo Truck. Seen here, this vehicle, Chevrolet Model G-7129, registration number 3114162, was photographed at Aberdeen Proving Ground on 5 March 1945.* National Archives

Below: *Another low-silhouette Chevrolet Model G-7129 Short-wheelbase Cargo Truck, registration number 3114161, is shown at Aberdeen Proving Ground on 5 March 1945. The driver's station was moved outboard, in line with the front left fender. A GMC engine was installed under the low-profile hood.* National Archives

Above: *The overall height of the Chevrolet G-7129 Short-wheelbase Cargo Truck was 99 inches, reducible to 67 inches by removing the cab top and tarpaulin, compared to an overall height of 106.5 inches for the G-506 Chevrolet 1 1/2-ton 4x4 Cargo Truck. The fuel tank was located on the left side of the chassis frame, to the front of which was a step for the driver. National Archives*

Below: *Chevrolet Model G-7129 registration number 3114161 is shown with the tarpaulin and cab top removed. Two spare tires were stored horizontally between the front seats. To the front of each of the rear fenders was a holder for a five-gallon liquid container. The tailpipe was to the front of the dual rear tires. National Archives*

Above: *The right side of Chevrolet Model G-7129 registration number 3114162 is shown with the tarpaulin and the cab top installed in a photo dated 5 March 1945. Along the bottom of the body and the cab between the wheels are, left to right, a storage box with a door with two latches and a piano hinge on the bottom, the battery; and a step made of bent metal rod.* National Archives

Below: *Registration number 3114161 is viewed from the right with the tarpaulin and cab top removed and the windshield lowered. The front of the chassis frame was redesigned for the Chevrolet Model G-7129: note the flared front end of the frame where it meets the front bumper.* National Archives

This rear view of Chevrolet Model G-7129 registration number 3114162 emphasizes the squat appearance of the low-profile 1 1/2-ton 4x4 trucks. The low-profile truck lacked the recessed toolbox below the tailgate. Taillights, bumperettes, and tow pintle were below the tailgate. National Archives

The rear of Chevrolet Model G-7129 registration number 3114161 is depicted with the tarpaulin removed. Despite the low height of the cargo-body floor, the crew steps were retained on the top of the tailgate. National Archives

was directed to obtain various styles and makes of the proposed low profile vehicles.

In early 1943 testing began at Holabird on many of the low profile trucks, including Chevrolet's offering, given Chevrolet model number G-7129. Like several of the low-profile prototypes, the driver was positioned outboard of the frame rail, in the area normally taken by the left running board.

Rather than the standard Chevrolet 235 cubic inch engine, the low profile truck was powered by the same GMC 270-cubic engine that powered the CCKW, coupled to a Clark transmission, also as used in the CCKW.

The Chevrolet, like the other low-profile trucks present, were not generally well received. Those inspecting the vehicles expressed overwhelming concerns that the compact design would make maintenance difficult, and the unconventional arrangement of the vehicles was uncomfortable.

Two months later, on 24 April 1943 H.H. Howard at the Tank-Automotive Center sent a memo to Brigadier General John K. Christmas providing an assessment of the low profile program. It read, "In the meeting of April 20th with your Engineering- Manufacturing Council consisting of Messrs. McCuen, Sheldrick, and Skelton, there was discussed the low silhouette motor transport vehicle program.

"It is the opinion of the above-named gentlemen that the program was impractical industrially on account of vast changes to plant, tooling, and parts required, and it is the further opinion of the Ordnance engineers present that the program had little merit to recommend it from the standpoint of utility of the vehicles proposed.

"Accordingly, it was the sense of the meeting that the low silhouette program should be killed off and that the vehicles should be withdrawn from circulation or from test in order to prevent development of any local enthusiasm with regard to one type or other."

General Christmas forwarded Howard's memo to Colonel Van Deusen, with the handwritten notation "Can't we close this project PDQ?"

On 21 July Van Deusen responded stating that the project was being cancelled and shipping orders were being issued to transfer all the low silhouette test vehicles to Aberdeen Proving Ground.

The conclusion of a 20 March 1944 report on the project was, "The trucks procured and tested under this project have been found to be unsuitable for procurement for military use. The designs tested do not possess sufficient advantage over present types of military vehicles to warrant adoption in their present form." The same report recommended: "That further development of low silhouette vehicles be terminated" and "That vehicles procured of under this project be disposed of in accordance with distribution chart dated 25 November 1943."

The same low-profile cargo truck, 3114161, is viewed from the upper rear, showing the rear wheel wells and the raised, single-occupant seat at the center of the front rack. There were two windshields, seen here in the lowered positions, with a snap-on canvas panel with two flexible-plastic windows spanning the gap between the two windshields. National Archives

Above: *Visible in this frontal view of a Chevrolet Model G-7129 is the scooped-out top of the bumper, above which are the winch-cable rollers. The so-called "snake" grille was formed from metal rods bent accordion style, with a diagonal brace on each side. Simplified brush guards for the headlights were provided. To the outboard side of the right brush guard was a square plate that appears to have been a placard holder. An identical fitting was installed between the left taillight and the left bumperette. National Archives*

Below: *The front winch and winch rollers, the curved shape of the front bumper, and details of the grille, the hood, the fenders, the headlights, and the brush guards of the Chevrolet Model G-7129 Low-profile Cargo Truck are highlighted in this frontal view. Wrapped around the tow hooks on the bumper is the leader chain of the winch cable. National Archives*

Chapter 13

Airborne Trucks

The development and use of "airborne" trucks by the US military during WWII is not widely known, and among those that are aware of this, most are familiar only with the airborne CCKW. Additionally, there were airborne variants of some Dodge trucks and of the 1½-ton 4x4 Chevrolet.

"Airborne" as it applies to any of these vehicles actually means air-transportable, not air-dropped. The earliest documented effort to develop such vehicles by the US Army was in 1943. 1st Lt Joe Berman and the men of the 2479th Quartermaster Truck Company, under 1st Lt Ron Bagley based in Port Moresby, Australia, conceived of a means to transport CCKW cargo trucks and Dodge weapons carriers in sections inside of standard C-47A transport aircraft. Inspired by those successes, in November 1943 Ordnance endorsed efforts to refine the concept in the States. The Office of the Chief of Ordnance – Detroit worked with the truck manufacturers to develop frame-splicing kits which would allow the vehicles to be loaded into the Douglas C-47A transport aircraft, which had a relatively small cargo door.

Kits were developed to splice the frames of both the cargo truck and the M6 bomb service truck. A common frame splicing kit was developed for both trucks, while the cargo body splicing kit used on the Chevrolet was common with the CCKW body splicing kit.

Because of the clearances involved, and the fact that all Chevrolet cargo trucks featured a closed cab, part of the modification process was the removal of the cab rear and roof from the top of the windshield to the belt line above the seatback. The windshield itself was cut off, but

From February to April 1944, the Ordnance Research Center at Aberdeen Proving Ground pursued Project 3297: a series of feasibility tests of airborne trucks of various models. The methodology of these tests was to break the vehicles down into smaller components to make them air-transportable. This included cutting the chassis in half and adding splicing kits to them, to allow them to be quickly disassembled prior to transport and reassembled after air-delivery. One of the vehicles tested was Chevrolet M6 Bomb Service Truck registration number 0038610, shown at Aberdeen on 27 March 1944 after the splicing kit had been installed. The right chassis splice is visible on the frame below the storage box with the white star on it. National Archives

Above: *In a 13 March 1944 photo, the seat assembly is being removed in advance of installing the splicing kit on the chassis frame. Note the bent mounting flanges on the bottoms of the seat supports.* National Archives

Below: *The crane assembly is being removed from the M6 Bomb Service Truck preparatory to installing the splicing kit on the chassis frame on 13 March 1944.* National Archives

Above: *The bed of the M6 Bomb Service Truck is being prepared for removal in the shop at Aberdeen Proving Ground on 13 March 1944.* National Archives

Below: *In this 21 March 1944 photo, the bed has been removed from the M6 Bomb Service Truck. On the rear of the right seat is a holder for a lubrication chart. Chalked on the side of the chassis frame is an arrow pointing to the bolt holes for the running board bracket.* National Archives

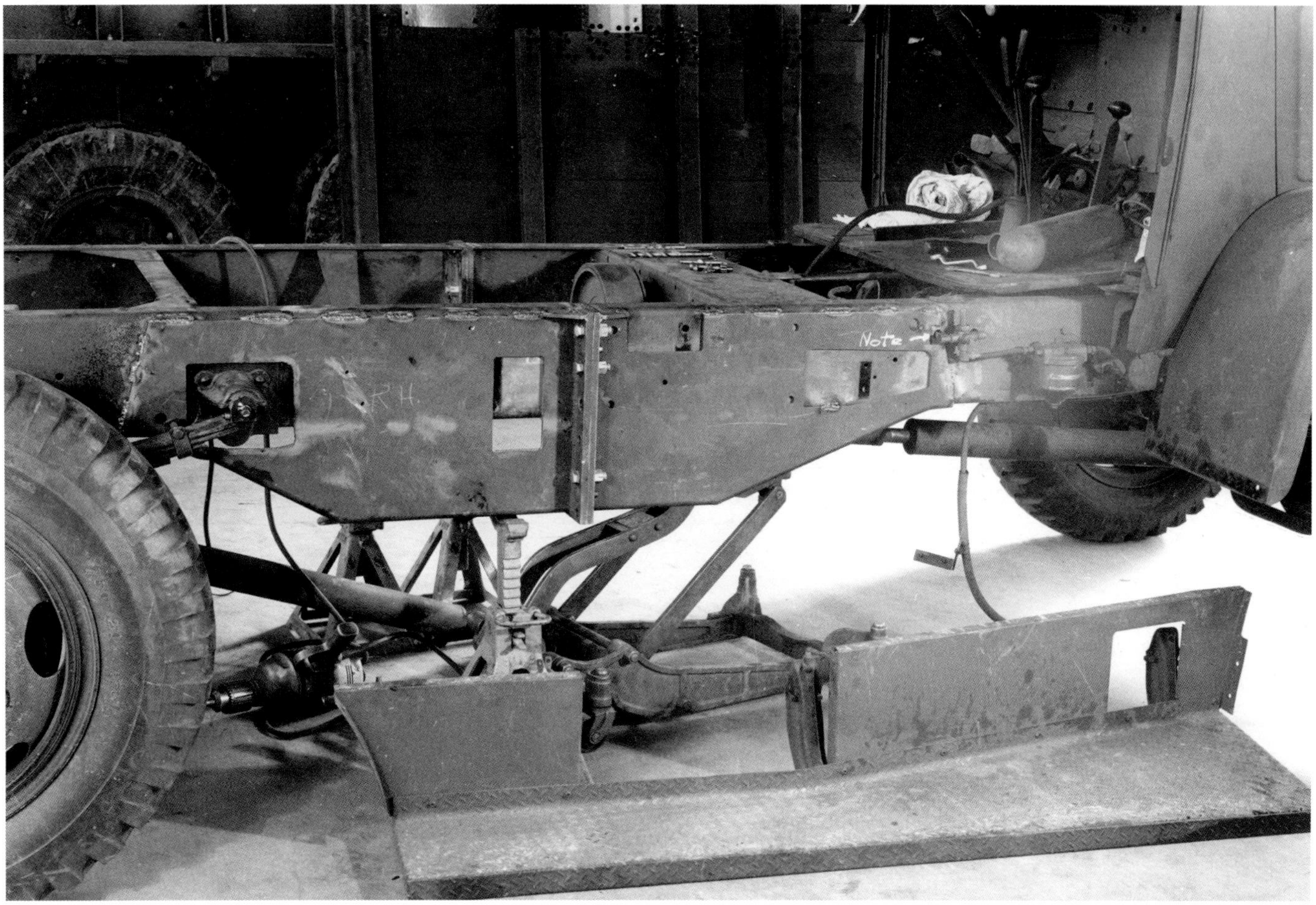

After the chassis frame has been cut on each side, front and rear splice plates have been welded to each side of the frame. Manufactured by the Chevrolet Truck Company, the splice plates provided a reinforced, flanged structure for joining the two halves of the chassis frame. The right splice plates are shown here, temporarily secured together with six hex screws and hex locking nuts. Cutouts in the splice plates provide clearance for the hanger for the rear spring and for mounting brackets. National Archives

unlike the cab roof which was discarded, brackets were provided to reinstall the windshield after delivery. The M6, lacking a closed cab, required only the removal of the windshield.

Testing of the frame splicing kit on the 1½-ton cargo truck began at Aberdeen Proving Ground on 12 April 1944. Chevrolet furnished the kit, as well as their drawing #3678602, for the project.

Beyond cutting the frame and installing the splicing plates, the brake lines had to be cut, but cutoff valves were provided so as to prevent the need to bleed the brakes in the field. The front bumper was removed, along with both front fenders and running boards. Because the front half of the truck, with its rear section supported by a dolly, was driven into the aircraft, a line was provided to allow the use of a 5-gallon gas can as a fuel tank. This was necessary because the truck's gas tank was not attached to the forward section of the vehicle.

Using these kits, two trucks could be transported in three C-47A aircraft, with the front and rear chassis sections of a single truck being loaded into one aircraft, and the body and various parts of a single truck filling half of a second aircraft.

The testing of the bomb service truck began a week prior to the testing of the cargo truck. Initially it was believed that due to its shorter 125-inch wheelbase, it would not be necessary to splice the frame of the M6, and thus no kit was provided. The truck was stripped, even of its rear axle, but tests found it was still too long to maneuver through the C-47A door. Accordingly, technicians at Aberdeen adapted a cargo truck splicing kit for the bomb truck. Based on their input, a common kit was developed for both types of vehicles.

In order to use the early cargo truck splicing kit on the M6 it was necessary to drill a couple of holes in the left front splicing plate to clear rivets, and for a notch

Above: *The cab seats, the rear part of the cab floor, the cab-cover support bow, and the running boards have been reinstalled on the chassis after the splice kit has been installed, on 27 March 1944. The splicing kit for the M6 was a modified version of the one employed on the Chevrolet 1 1/2-ton 4x4 Airborne Cargo Truck.* National Archives

Below: *In another photo taken on 27 March 1944, the crane has been reinstalled on Chevrolet M6 Bomb Service Truck registration number 0038610. The splice on the chassis frame is visible below the storage box, with all of the hex screws and nuts installed.* National Archives

Above: *During Project 3297/8-1-7, the aircraft-loading phase of the project, workers have begun disassembly of Chevrolet M6 Bomb Service Truck 0038610 at Aberdeen Proving Ground on 28 March 1944. The vehicle would be broken down into its main components for loading into a Douglas C-47A Skytrain.* National Archives

Below: *The front half of the M6 Bomb Service Truck has been detached from the rear half. The rear of the section is being supported by pairs of hydraulic jacks and ratchet jacks, and a single-wheeled dolly has been installed. This dolly made it possible for the front section to be moved around under its own power. Attached to the back of the passenger seat is the temporary fuel tank to allow the vehicle to be driven onto the aircraft.* National Archives

Above: *The rear section of the M6 Bomb Service Truck is viewed from the front left, with a clear view of the mounting flange of the left rear splice plate. The front of the rear drive shaft is lying on the ground. The fuel tank and the rear axle also are in view.* National Archives

Below: *The disassembled M6 Bomb Service Truck is ready for loading into a Douglas C-47A Skytrain at the Aberdeen Proving Ground Airport on 28 March 1944. From left to right are the forward section of the chassis, the detached dolly next to the rear section of the chassis, the front bumper, fenders, seats and cab bow, storage boxes, crane with windshield leaning against it, and bed. Tests at Aberdeen showed that it took a six-man crew about one and a half hours to disassemble the truck and two hours to reassemble it.* National Archives

to be cut in the right front splicing plate. It was also necessary to relocate the truck's battery box.

After the frame splicing kit was installed, a study determined that a six-man crew could disassemble the bomb service truck in 1½-hours, and reassemble it in 2 hours.

While air-transportable versions of the Dodge WC-51 and CCKW were produced by their respective factories, as well as contractors, and are so noted in "Engineering of Transport Vehicles," Chief of Ordnance, Detroit, supplement for period 1 October 1944 to 31 May 1945, the same document notes that "Truck, 1-1/2 Ton, 4x4, Bomb Service, M6, may be modified without difficulty, for airborne transportation, and for this vehicle also, kits for field modification are under procurement." No mention of air transport of the cargo variant is made.

Above: *Although an 8 April 1944 report of tests of the M6 Airborne Bomb Service Truck at Aberdeen Proving Ground stated that as of that date the vehicle had not yet been test-loaded in to a C-47 Skytrain, this photograph from Abderdeen dated 3 March 1944 shows that an attempt had been made before 8 April to load the vehicle into a Skytrain. In this test, the front and rear sections of the chassis were joined together, but the windshield, the cab bow, the rear axle and wheels, the bed, the seats, and the crane were removed, and a one-wheeled dolly supported the rear of the chassis. Even by removing these components, the vehicle still could not fit into the C-47A.* National Archives

Below: *In addition to the Chevrolet M6 Bomb Service Truck, the Chevrolet 1 1/2-ton 4x4 Cargo Truck was the subject of tests of the feasibility of airborne vehicles. This example, registration number 393413, was prepared for disassembly when photographed in a shop at Aberdeen Proving Ground on 30 March 1944. Some of the major modifications to enable the truck to be disassembled and loaded in a cargo plane included a splice kit for the chassis frame, the removal of the cab roof, and a removable windshield with splices at the level of the top of the cowl.* National Archives

Above: *The following series of photos was taken at Aberdeen Proving Ground on 21 February 1944 during the conversion of Chevrolet 1 1/2-ton 4x4 Cargo Truck registration number 292413 to an airborne vehicle. In this photo, the cargo body, left, has been removed from the truck chassis, right. Work has not yet begun on the chassis modifications.* National Archives

Below: *The six rivets, three vertical ones per side, have been removed from the front bumper.* National Archives

Above: *At this point in the preparation of the cargo truck for conversion to an airborne truck, the front fenders and bumper, running boards and aprons and their brackets, the battery box, and the spare tire had been removed from the chassis.* National Archives

Below: *The fuel tank is being prepared for removal from the chassis. The view is from the left side, with the exhaust and the spare-tire carrier and its locking handle in the foreground.* National Archives

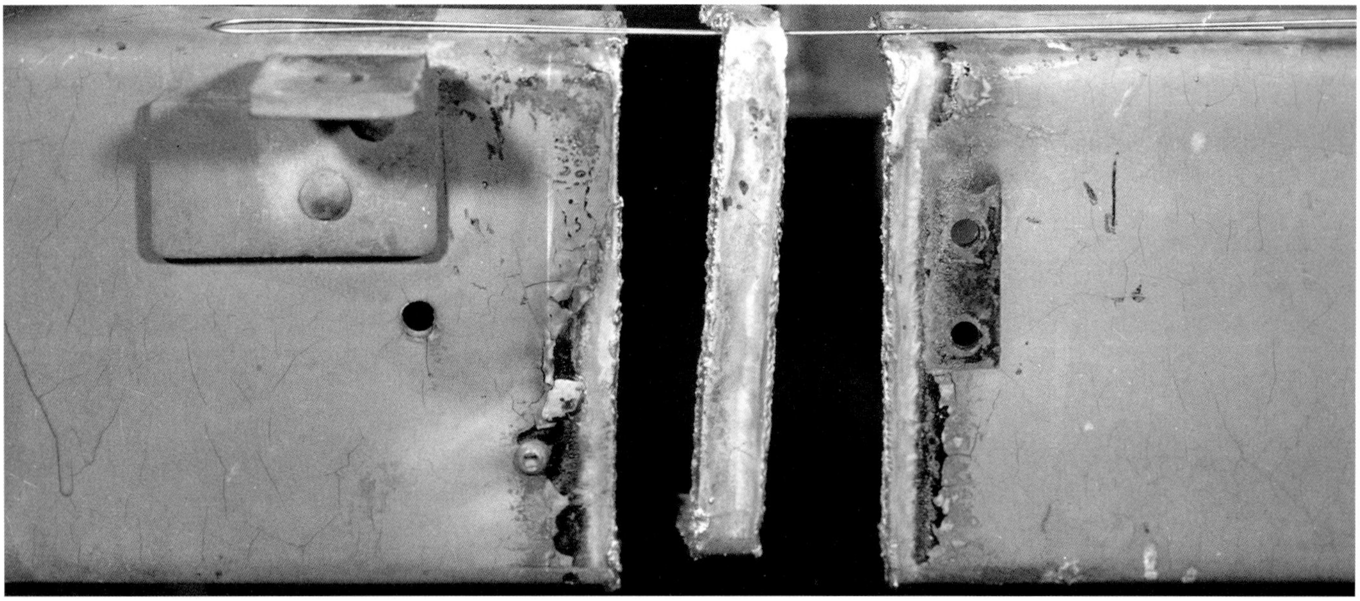

Above: *To install the chassis splice, the frame was divided by a cutting torch on each side to the rear of the cab, and one inch of material was removed. This photo was taken on the left side.* National Archives

Below: *The components of the splice kit for the Chevrolet 1 1/2-ton 4x4 Cargo Truck are laid out at Aberdeen Proving Ground on 21 February 1944. The two side plates are in the background, with flanges secured with hex screws and nuts, and cutouts for fittings on the sides of the chassis frame. In the foreground are a small tank, a flexible fuel line and shutoff valve to install where the fuel line was cut, and hydraulic lines and valves for where the brake lines were severed. One valve controlled the hydraulic fluid in the rear section and the other controlled the fluid in the front section. A small fuel tank, intended to hold only enough fuel to load and unload the truck, is in the left foreground.* National Archives

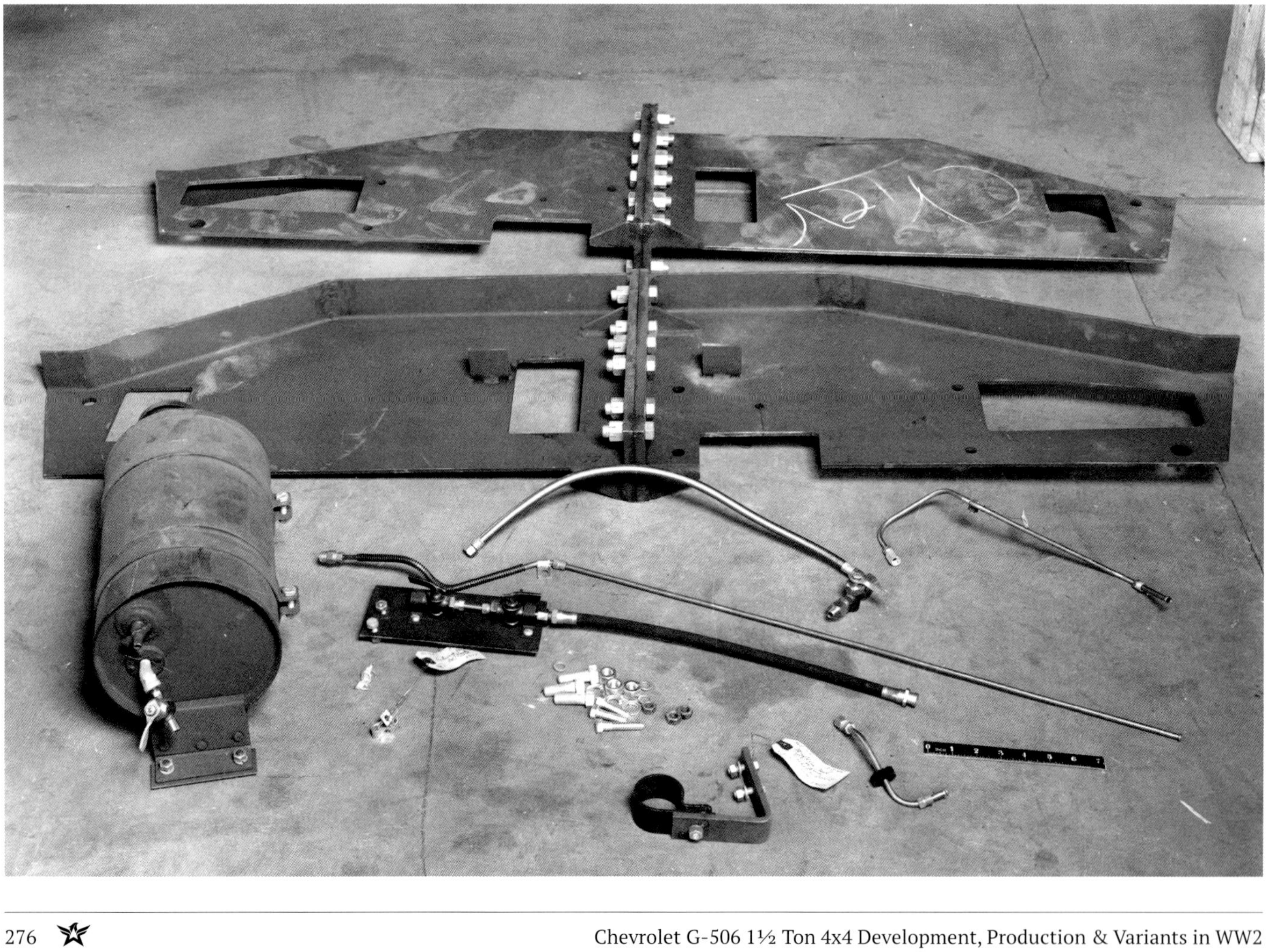

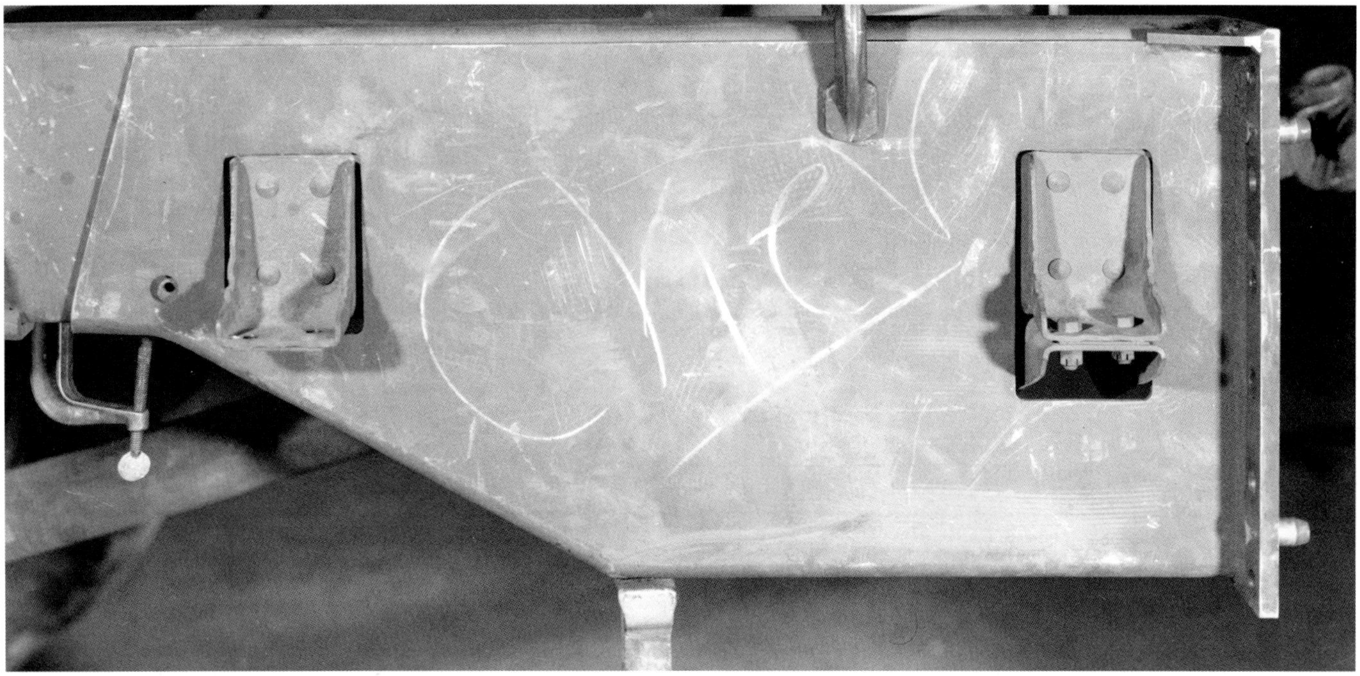

Above: *The right-hand splice plate is clamped to the rear section of the chassis frame and is ready for welding. Protruding through the pre-cut openings in the plate are the brackets for the fuel-tank hangers. To the front of the plate is a flange, with two locating studs, which formed a fastening surface for the front splice plate.* National Archives

Below: *The rear of the front section of the truck has been lifted by chain hoist, showing the rear faces of the flanges of the front splice plates, as installed on the chassis frame. Two locating studs are on the left flange but not the right; the locating studs were reversed on the flanges of the rear splice plates. A good view is available of components on the underside of the truck, including the front axle and drive shaft, the transfer case, the exhaust, and the muffler.* National Archives

Above: *In another 24 February 1944 photograph, a jack is supporting the rear section of the airborne cargo truck, the drive shaft resting on the floor; both of the rear splice plates have been welded to the frame. Short weld beads are visible at intervals on the top of the closer splice plate. Both of the outer dual rear wheels have been removed.* National Archives

Below: *A worker is using a torch to cut the roof off of the cab at Aberdeen on 24 February 1944. She is making the cut just above the horizontal frame member on the rear of the cab.* National Archives

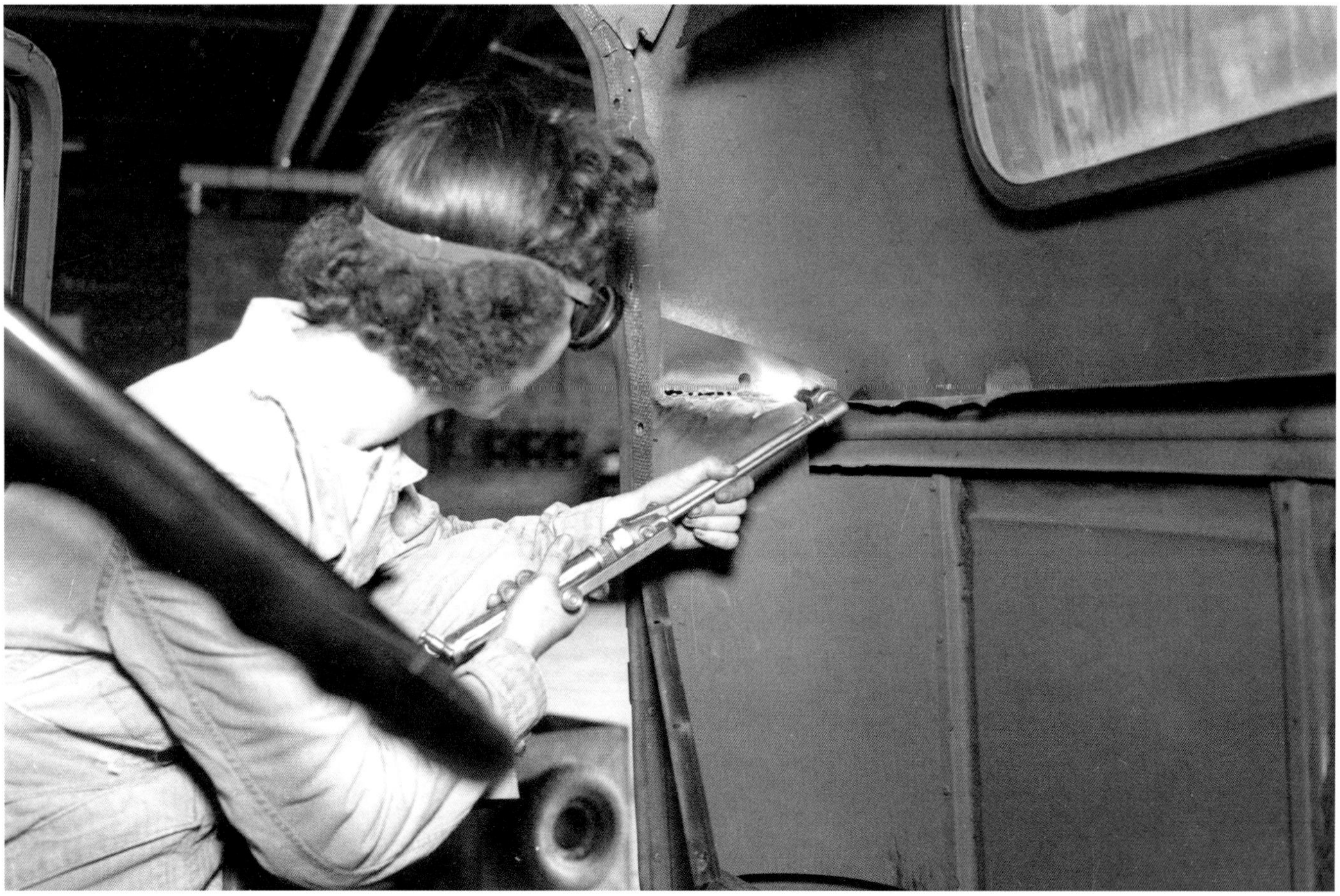

Above: *Also on 24 February 1944, another worker is cutting the metal of the right A pillar of the airborne cargo truck, during the removal of the roof. A similar cut already has been made on the lower part of the windshield frame on the door.* National Archives

Below: *The cargo body of Chevrolet 1 1/2-ton 4x4 Cargo Truck registration number 292413 is being prepared for cutting in half from side to side on 24 February 1944. Marks have been made for the cut to the front of the left center stake pocket in the foreground, and the vertical cut already has been made to the opposite side of the body. A steel straightedge is clamped alongside the cut, apparently as a burn guide for the cutting torch, seen here lying on the floor of the body.* National Archives

Above: *The battery and the modified battery box for the airborne cargo truck are the subject of this photograph. Under the door sill adjacent to the battery cable is the fuel filter. National Archives*

Below: *As seen in a 28 February 1944 photo at Aberdeen Proving Ground, the rear of the front section of the airborne cargo truck initially was supported by a dolly consisting of a rubber-tired wheel from a 1-ton trailer mounted on a section of angle iron. This enabled the front section to move under its own power when the airborne cargo truck was disassembled for air transport. Later, a universal dolly was developed, employing the wheel and fork assembly from the three-wheeled bomb loader produced by the Weaver Manufacturing Company. National Archives*

The cargo body of a Chevrolet 1 1/2-ton 4x4 Airborne Cargo Truck is resting upright on its right side, the tailgate to the left. The body sides and bed have been cut laterally in half, and holes have been drilled on the bed of the forward section for attaching the splicing angle iron, called the tie angle. The cut on the bed was along the edge of the center bolster. The tie angle on the rear of the forward section of the body was welded, not bolted, in place. National Archives

Above: *A worker at Aberdeen Proving Ground points to the installation of the tie angle on the forward edge of the rear body section of airborne cargo truck 393413 on 12 April 1944. The tie rod was face-bolted to the center bolster of the body.* National Archives

Below: *Another view dated 12 April 1944 shows the installation of support plates on the body and body-sill splices of Chevrolet 1 1/2-ton 4x4 Airborne Cargo Truck 393413. At this time the truck was the subject of to Ordnance Research Center Project 3297/8-1-7, which entailed tests of loading the vehicle on a transport plane at Aberdeen Proving Ground Airport.* National Archives

Above: *For the loading test, load no. 1 consisted of the front section and the rear section of the chassis, while load no. 2 included the two sections of the body, stakes and racks, windshield, three spare tires, fenders, front bumper, running boards, and other detached components. Load no. 2 is shown in this photograph from 12 April 1944.* National Archives

Below: *Load no. 1 from the Project 3297/8-1-7 aircraft-loading test is shown in the shop at Aberdeen Proving Ground on 30 March 1944. The load consisted of the disassembled front and rear sections of the chassis. Jacks are supporting the front of the rear section, and the dolly is installed on the rear of the cab, the wheel resting on a wooden block.* National Archives

Above: *As viewed from the rear, the front section of the chassis of the airborne cargo truck is secured in the forward end of the cargo compartment of a C-47A Skytrain at Aberdeen Proving Ground Airport on 12 April 1944. When entering the aircraft, the right corner of the cab failed to clear the door, so the front spring shackle was removed, dropping that corner of the chassis enough to allow the vehicle to enter the aircraft.* National Archives

Below: *The rear section of the airborne cargo truck has been secured in the C-47A Skytrain behind the front section of the chassis. The front of the rear section is closest to the camera.* National Archives

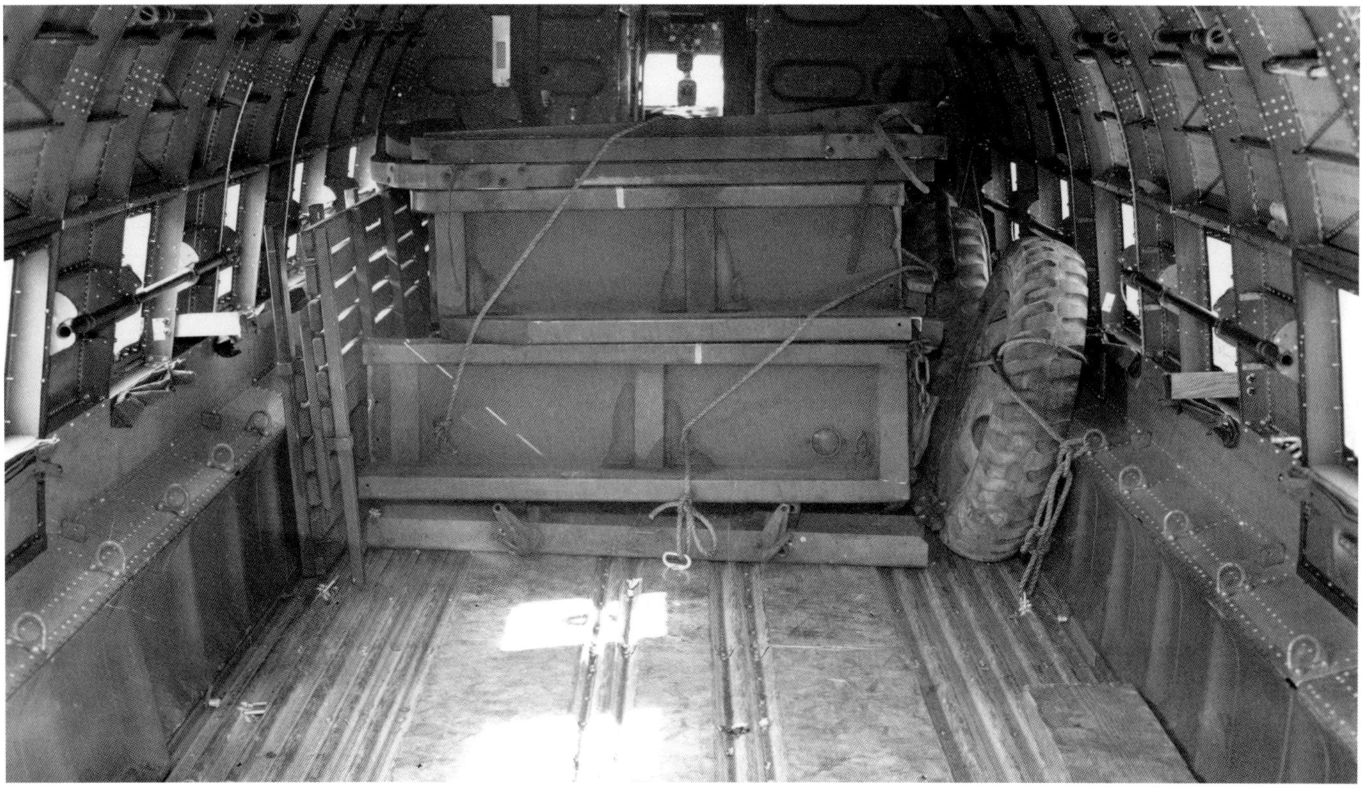

Above: *Load no. 2 is stored in the forward end of the C-47A on 3 March 1944. The two sections of the body are stacked; to the right are two of the tires, and to the left are the stakes and racks. Project 3297/8-1-7 determined that two of the No. 2 loads could be carried in a single C-47 Skytrain.* National Archives

Below: *Chevrolet 1 1/2-ton 4x4 Airborne Cargo Truck registration number 3372393 is seen at Aberdeen Proving Ground on 19 December 1944. This vehicle has the wooden body, and a splice plate is visible to the immediate front of the center stake pocket. The flanged splice on the left A-pillar of the windshield is clearly visible. The service headlights have been dismounted.* Ordnance Museum

Chapter 14

Auto-Railer Conversions

The lack of major highway transport arteries in China during WWII and the creation of the Ledo (Stillwell) Road to remedy this is well known. While the primitive state of most Chinese roads at that time was widely publicized, less well known is the existence of an extensive rail network in the nation.

In the United States, the gauge, or distance between the rails of most railroads is 4 feet 8-1/2-inches, and this is known as standard gauge. However, other gauges of track do exist, particularly overseas. In WWII China, there was a considerable amount of 42-inch gauge track in use. Allied strategists felt that if the trucks could be made to operate on the railroad a considerable amount of labor, and more importantly, time, could be saved by eliminating the transloading of cargo between train and truck.

The Evans Products Company of Detroit, Michigan began producing conversion kits that allowed highway

In early 1945 the Ordnance Research and Development Center, Aberdeen Proving Ground, Maryland, undertook Project No. 6209, to test a kit for modifying several types of military vehicles for railroad operation, including the Chevrolet 1 1/2-ton 4x4. This type of conversion, known as Auto-Railer, entailed fitting steel, flanged pilot wheels on a truck to allow it to operate on tracks under its own power. Auto-Railers were especially needed in China, where railroad lines often served remote places inaccessible by roads. The Evans Products Company, Auto-Railer Division, Detroit, Michigan, produced the Auto-Railer kits for the Office, Chief of Transportation. Here, Chevrolet K-44 registration number 0092413 is undergoing tests of the auto-railer kit at Aberdeen on 9 June 1945. National Archives

Above: *The front and rear steel-flange pilot wheels were on frames that were lowered for operation on railroad tracks and raised for operation off the tracks, since the auto-railer vehicles were intended to be able to operate on or off tracks. The Auto-Railer nomenclature plate is on the front axle. The front tires straddled the tracks during rail operation and did not provide propelling power.* National Archives

Below: *In order for the inner rear tires, which propelled the vehicle on tracks, to gain traction on the railroad tracks, it was necessary to reduce their tread from 57 inches to 43 3/16 inches. This was done by trimming down the outer ends of the axle shafts and the differential carrier. The pilot wheels and axles were raised and lowered hydraulically.* National Archives

Another modification that was made to the Chevrolet K-44 was to narrow the chassis frame from the rear springs aft, from 40 ¾-inches to 22 1/4-inches. This involved cutting the frame and making a splice to the immediate front of the front hangers of the rear springs. The shortened rear cross member of the chassis frame is visible below the center of the body. National Archives

vehicles to be operated on railroad tracks in the 1930s. Normally, this conversion was done by adding flanged railroad wheels just ahead of the vehicle's standard front wheels, and just behind the vehicle's normal rear wheels. The railroad wheels could be raised or lowered by means of a jackscrew or hydraulics. The vehicle was propelled on rail by the normal drive wheels contacting the top of the rail. In order to align the truck tires with the railroad track, sometimes the axle width or wheel offset had to be modified. Evans dubbed their product the Auto-Railer.

Ordnance asked Evans to develop Auto-Railer systems that would convert certain military vehicles for use on 42-inch gauge railway track. The vehicles that Evans was asked to convert were the ¼-ton Jeep, the Dodge 3/4-ton 4x4 and the 1½ -ton Chevrolet 4x4. Later a GMC CCKW 2 1/2-ton 6x6 was converted as well.

While the jeep was adapted by merely adding the rail kit and using different wheels, reducing the vehicle track from 49 to 45 inches, the Chevrolet required considerably more work. While the front axle was not altered, at the rear the Chevrolet rear axle tread, normally 57 inches for the inside dual tires, had to be reduced to 43 3/16-inches for rail use. This was done by removing a section from the rear axle tube on either side of the differential carrier and then shortening the axle shafts. To provide clearance for the narrowed wheel tread it was further required that the frame be narrowed from 40 ¾-inches to 18 1/2-inches. This required cutting the frame and making a splice just ahead of the forward rear

Above: *Chevrolet K-44 registration number 0092413 is viewed from the upper right during tests of the Auto-Railer kit at Aberdeen Proving Ground on 9 June 1945. The right rear wheel is visible to the rear of the rear fender, and the front wheel and part of the front axle are also in view.* National Archives

Below: *On 29 June 1945, K-44 0092413 has been maneuvering through thick mud while being subjected to a road test on the Perryman cross-country course at Aberdeen. The front rail wheels were striking ground during this operation.* National Archives

The K-44 with the Auto-Railer kit is viewed from the right side while maneuvering through thick mud at Aberdeen Proving Ground on 29 June 1945. The complete Auto-Railer kit for the 1 1/2-ton 4x4 weighted 1,480 pounds. National Archives

spring hangers. No changes were made to the front axle or front frame width, as the front of the vehicle would rest entirely on the rail wheels, the rear tires providing the only propulsion or braking for the vehicle.

Preliminary tests were made at the Evans plant in Detroit on 300 feet of specially prepared track as well as on city streets and in a field. Those test showed that the rail operation of all three vehicles were satisfactory, but that only the jeep, which had all four rubber wheels contracting the rails, had adequate braking.

The test vehicle for the Chevrolet-based conversion was K-44 earthboring and polesetting truck registration number 0092413, serial number 9NR37-1989.

The Aberdeen test report dated 1 August 1945 covering all three vehicles concluded "As a result of test operation to date, it is concluded that the Auto-Railer vehicles are unsatisfactory for cross country operation due to their small angle of approach and departure and low ground clearance of rail pilot wheel and axle assemblies in retracted position."

During the cross-country tests on the Perryman course on 29 June 1945, the rear rail wheels also tended to strike the ground. The performance of the Auto-Railer kit during road and cross-country driving was important, since it was anticipated that these trucks would be able to operate as freely on those surfaces as on railroad tracks. National Archives

Chapter 15

The Chevrolet in the Field

Production of the 1½-ton 4x4 Chevrolet began in 1940, with the intent at that time that it be one of the five most crucial types of vehicles of the US Army. Once the United States became directly involved in WWII, other vehicles, most notably the GMC CCKW and the Dodge WC-62/63, began to erode the role of the Chevy.

Photographs of the trucks in service with US troops are abundant in the pre and early war years, and lessen dramatically as the war wore on.

The exceptions to this are the assorted specialized trucks - the bomb service, gunnery trainers, fire trucks and various Signal Corps vehicles – these vehicles pushed on for the duration of the war in widespread US service.

However, as we have seen, the cargo and dump trucks continued to pour from the Chevrolet assembly lines. Chevrolet built over 150,000 of the 1½-ton 4x4 trucks, including 113,077 cargo trucks. Where did all those cargo trucks go if not seen in photos of US troops? To our allies

Members of Company A, 107th Engineers, cross a 10-15-ton bridge that they had erected in 80 minutes during maneuvers at Camp Livingston, Louisiana, on 23 May 1941. They are seated in 1941 Chevrolet 1 1/2-ton 4x4 Dump Trucks: registration numbers W-331423, left, and W-331408, right. Both vehicles have the early-style Hercules dump bodies with ledges on the sides supported by triangular braces. Unit markings are faintly visible on the doors. National Archives

Above: *Vehicles, including two Chevrolet 1 1/2-ton 4x4 Cargo Trucks assigned to Troop A, 8th Engineer Battalion, are filing across a pontoon bridge constructed by Troop B of the 8th Engineers during Army maneuvers in May 1941. Storage boxes are secured to the ledges alongside the cargo bodies. The vehicle to the front is a 1941-production truck, registration number W-324592.* National Archives

Below: *Five 1941-model Chevrolet 1 1/2-ton 4x4 Cargo Trucks assigned to a Signal Corps detachment are lined up at Fort Monmouth, New Jersey, in June 1941. To the rear is a civilian-type vehicle serving as a cargo truck. The nearest two vehicles bear registration numbers W-320592 and W-330668, produced under Quartermaster contract W-398-QM-8278.* National Archives

Above: *The same column of trucks and soldiers seen in the preceding photo are viewed from a different perspective on the same date, at Fort Monmouth in June 1941. Although only faintly visible in these two photos, "FRONT" is stenciled in black on the upper fronts of the tarpaulins.* National Archives

Below: *Members of a Signal Corps field-telephone section are setting up a switchboard next to a Chevrolet 1 1/2-ton 4x4 Cargo Truck during an exercise at Fort Monmouth, New Jersey, around mid-July 1941. The sergeant seated on the chest is operating a BD-72 12-line switchboard, while the soldier to the far right is using an EE-65 test set.* National Archives

Above: *A mobile switchboard unit mounted on a Chevrolet 1 1/2-ton 4x4 Cargo Truck was assigned to the 62nd Signal Battalion, based at Fort McPherson, Georgia, in the summer of 1941. This photo emphasizes the terminal block on the left side of the body. A hinged wooden cover for the terminal block is swung open.* National Archives

Below: *The mobile switchboard of the 62nd Signal Battalion is viewed from the rear with the wooden rear doors open, showing the two-place switchboard located inside. Swivel seats are supplied for the operators. Stenciled in white on the left mudguard is "62nd SIG BN 16."* National Archives

This is a closer view of the two-place switchboard inside the mobile switchboard truck serving with the 62nd Signal Battalion in 1941. The interiors of the rear doors, visible to the sides, were made of wood and were painted or stained a dark color, likely Olive Drab. The doors had Z-bracing. The sides of the enclosure appear to have been made of lumber painted in a light color. A conical light fixture was attached to the ceiling. National Archives

is the answer. The British Empire got over 2,500; but the number the Soviet Union received is in the neighborhood of 60,000. The uncertainty is a result of two things. First, reports separating 1½-ton 4x4 trucks by manufacturer have not surfaced, and secondly, two different numbers are reflected in official reports. One report indicates that the Soviets received 61,349 trucks in this class, another gives the number as 60,475. Similarly, the UK number is given as 2,730 and 2,731 respectively. In any event, it is clear that the Soviets received the bulk of the cargo trucks. As pointed out earlier, the high Soviet demand for the vehicles is the reason these trucks retained their closed cab for the duration of the war.

Regrettably, photographs of wartime Soviet use of Lend-Lease vehicles remain difficult to obtain.

By February 1953, many models of the Chevrolet had been classified as Obsolete or Substitute Standard by the US military. Only a few years later they all were.

Above: *A multitude of trucks, including many G-506 Chevrolets, is stored at the Brooklyn Army Base, Brooklyn, New York, in August 1941, awaiting shipment overseas. The row of vehicles to the left are Chevrolet 1 1/2-ton 4x4 Cargo Trucks, with registration number W-332595 at the front.* National Archives

Below: *Vehicles assigned to the Signal Corps and the Army Air Corps, including trucks, tanker trailers, and special-purpose trailers, are awaiting shipment overseas at the Brooklyn Army Base in August 1941. The group of five trucks to the right are Signal Corps vehicles that transported the components of the SCR-270 Early-Warning Radar. The second truck from the right is a Chevrolet K-33 Stake Body Truck, based on the 1 1/2-ton 4x4 cab-over-engine truck. This vehicle towed the K-22 Antenna Trailer, which held the antenna array of the SCR-270. Flanking the K-33 are two Autocar K-32 Prime Movers, and the Autocar K-31 and K-30 vans containing the power plant and the operating equipment for the SCR-270 are the fourth and fifth vehicles from the right.* National Archives

Above: *While participating in the First Army Maneuvers in the Carolinas in the autumn of 1941, members of Battery E, 209th Coast Artillery (Antiaircraft), take a break on their Chevrolet 1 1/2-ton 4x4 Cargo Trucks prior to a rapid movement from their base near Hoffman, North Carolina. The trucks are equipped with water-cooled Browning M2 .50-caliber machine guns on pedestal mounts. The truck in the foreground, registration number W-334587, has a pioneer tool rack on the side of the body.* National Archives

Below: *Behind the steering wheel of this 1941 Chevrolet 1 1/2-ton 4x4 Cargo Truck, W-330225, at Fort Benning, Georgia, in October 1941 is Pvt. F. K. Kurtz of the 161st Signal Photo Company. Painted on the door is "161st SIG PHOTO CO."* National Archives

Above: Below: *A civilian man and a boy look on as troops go about their business around a row of Chevrolet 1 1/2-ton 4x4 Cargo Trucks during October 1941. The truck to the right and the next one from the right bear, respectively, registration numbers W-341947 and W-331454. Illegible unit markings are present on the bumpers and the cab doors.* Jim Gilmore collection

Below: *Elements of a mixed convoy are getting into position at Fort Benning, Georgia, in October 1941. In general, the trucks with five bows over the cargo compartments are G-506 Chevrolet Cargo Trucks. Also in evidence are several Dodge carryalls, command and reconnaissance cars, and half-ton trucks.* National Archives

Above: *This photograph was taken on 16 October 1941 to document a U.S. Army Signal Corps Division Assignment Unit and its equipment. The vehicle to the right is a Chevrolet 1 1/2-ton 4x4 Panel Delivery Truck, while the one behind it is a Dodge 1/2-ton Carryall with a wooden baggage rack on the roof.* National Archives

Below: *Photographic lab equipment and the individual equipment of the driver and an officer are loaded into the Chevrolet 1 1/2-ton 4x4 Panel Delivery Truck seen in the preceding photograph. A notation with the photograph explained that a boxed photographic enlarger was placed at the front of the compartment and almost reached the ceiling.* National Archives

Above: *The number-16 vehicle of the 4th Signal Company, 4th Division, served as a complete mobile switchboard based on Chevrolet 1 1/2-ton 4x4 Cargo Truck registration number W-323081, a 1941 G4112. At an encampment, a Signal Corpsman is rigging telephone wires over a frame made of pipe sections and fittings on the front of the cargo body. "4TH SIG CO 16" is stenciled on the bumper.* General Motors LLC

Below: *At an unidentified Army base during 1941, two rows of Chevrolet 1 1/2-ton 4x4 Cargo Trucks are hitched to trailers holding pontoons, for constructing a pontoon bridge. The trucks will deliver the pontoons to a river, where combat engineers will assemble a temporary bridge from these craft along with treadway sections.*

Above: *A group of G.I.s is loading supplies on a Chevrolet 1 1/2-ton 4x4 Cargo Truck in Iceland in 1941. These supplies were destined for a hospital. United States forces occupied Iceland in the summer of 1941 as part of an Allied plan to deny the use of that island in the North Atlantic to the Germans.* National Archives

Below: *This Chevrolet 4x4 Panel Delivery Truck at Fort Benning on 20 January 1942 has an unusual feature: a simple rack for a five-gallon liquid container fashioned from a bent piece of metal, attached to the front of the rear fender. This truck, registration number W-356761, has markings for the 161st Signal Photo Company on the cab door.* National Archives

Above: *Chevrolet 1 1/2-ton 4x4 Cargo Truck registration number 330179 was photographed during an exercise at Fort Benning, Georgia, on 20 January 1942. Of interest are the two five-gallon liquid containers on the ledge on the side of the cargo body, held in place by a unique rack fabricated from steel strapping material. To the front of these liquid containers, and partially obscured by the tarpaulin, is a pioneer tool rack, with no tools visible.* National Archives

Below: *Three Signal Corpsmen are winding uninsulated wire near a Chevrolet 1 1/2-ton 4x4 Telephone Maintenance Truck during an exercise in January 1942. The stored gin poles of the truck's derrick are protruding from the front of the body, alongside the driver's door. Rolls of uninsulated wire are stored on the inside of right side of the body.* National Archives

Above: *A member of Photo Unit 1 at Camp Funston, Kansas, demonstrates the storage provisions for photographic supplies in the rear of a Chevrolet 1 1/2-ton panel truck, on January 26, 1942. The rack was made of wood, some of which appears to be reused lumber, and was painted, evidently, Olive Drab. To the left of the rack is storage space for two five-gallon liquid containers and pioneer tools.* National Archives

Below: *Two dungaree-clad drivers with the 162nd Signal Photo Company are scrubbing their Chevrolet 1 1/2-ton 4x4 Cargo Trucks without winches on the wash rack at Camp Robinson, Arkansas, on 17 February 1942. Visible on each of the bumpers, and on the cab doors of the truck to the left, is the inscription, "162ND SIG PHOTO CO" within a rectangular border. The truck to the left is registration number W-351216, while the one to the right is W-345720.* National Archives

Above: *The U.S. Army occupied military bases on Iceland during World War II. Here, members of a U.S. Army signal battalion are preparing to bore holes for installing telephones using a 1941-model Chevrolet K-44 1 1/2-ton 4x4 Pole-setter Truck with Earth Borer.* National Archives

Below: *The same signal crew shown in the preceding photograph stand by as the telephone earth borer does its work. A clear view is available of the borer case, to the right of the soldier operating the equipment to the left. It was necessary to keep the auger at a low rate of speed—about 25 rpm—for the first 18 inches or so of depth, to avoid damage from rocks. After the first 18 inches, higher speeds were permitted.* National Archives

Above: *Drivers under a Lieutenant Bornholdt are assembled in front of their vehicles at the motor pool at Camp Robinson, Arkansas, on March 5, 1942. In addition to two jeeps and two Dodge G-505 1/2-ton 4x4 trucks to the left and a row of trailers and cargo trucks to the rear, there are five G-506 Chevrolet panel trucks to the right. Notably, these panel trucks are not K-51 radio trucks.* National Archives

Below: *A fire truck in U.S. Navy markings, left, and a Chevrolet Cab-over-engine Fire Truck, right, are parked at an unidentified base in August 1942. A placard marked "FIRE DEPT." is affixed to the grille of the cab-over truck. Two spotlights are on the front of the cab roof. Pinstriping is painted on the driver's door.* National Archives

Above: *The U.S. Army Signal Corps took this right side view of a Chevrolet K-51 at the Desert Training Center, Indio, California, on June 8, 1942, as an identification photo. Four whip antennas are installed on the roof. The registration number, 365643, is faintly visible on the hood.* National Archives

Below: *The same K-51 Panel Truck seen in the preceding photo is viewed from the rear at Indio. A measuring stick is leaning up against the vehicle, for reference. Note the light-colored protective material wrapped and secured around the cable reels.* National Archives

Above: *A Chevrolet M6 Bomb Service Truck of the 341st Bomb Group is being used to bring bombs to a U.S. Army Air Forces B-25 Mitchell medium bomber in India in September 1942. The men are rolling 500-pound bombs, minus the fins, off an M5 Bomb Trailer hitched to the truck.* National Archives

Below: *Ordnance men are bringing a load of 500-pound bombs to a B-25 with a Chevrolet M6 Bomb Service Truck and an M5 Bomb Trailer at an airbase in India in September 1942. Stickers on the windshield warn that the maximum speed is 20 miles per hour. Another trailer full of bombs is faintly visible in the left background. Ordnance men were tasked with transporting bombs to the aircraft, while armorers were the specialists who mounted the bombs on the bomb racks.* National Archives

Above: *In 1942 the U.S. Army Corps of Engineers constructed the Alaska Highway, also known as the ALCAN Highway, a supply route from the northwestern United States through western Canada to Alaska. Chevrolet G-506 trucks played a role in the building of this strategically important highway. Here, a barge is transporting a Chevrolet 1 1/2-ton 4x4 Dump Truck with a load of earth or gravel across a river during the construction of the highway around August 1942.* National Archives

Below: *A Chevrolet 1 1/2-ton 4x4 Dump Truck with winch, registration number 3104579, leads several other trucks across a ford in a stream next to a site where engineers are constructing a bridge from logs on the Alaska Highway during September 1942. The Chevrolet dump truck has chains on the front tires and on the dual rear tires for better traction on soft ground. "B-6" is stenciled on the left side of its bumper.* MVPA

Above: *A Chevrolet 1 1/2-ton 4x4 vehicle, either a cargo truck or a dump truck, has become mired along a muddy road during the construction of the Alaska Highway in September 1942. A placard with writing, albeit illegible, is affixed to the asymmetrical brush guard, and the side panel of the hood is missing.* MVPA

Below: *A mixed lot of U.S. military vehicles is being unloaded from railroad flatcars at Dawson Creek, British Columbia, in September 1942 during the final weeks of construction of the Alaska Highway. The first and second trucks on the ground in the foreground are Chevrolet G-506 vehicles. The registration number on the hood of the nearest vehicle appears to be W-348781, which pertained to a cargo truck.* MVPA

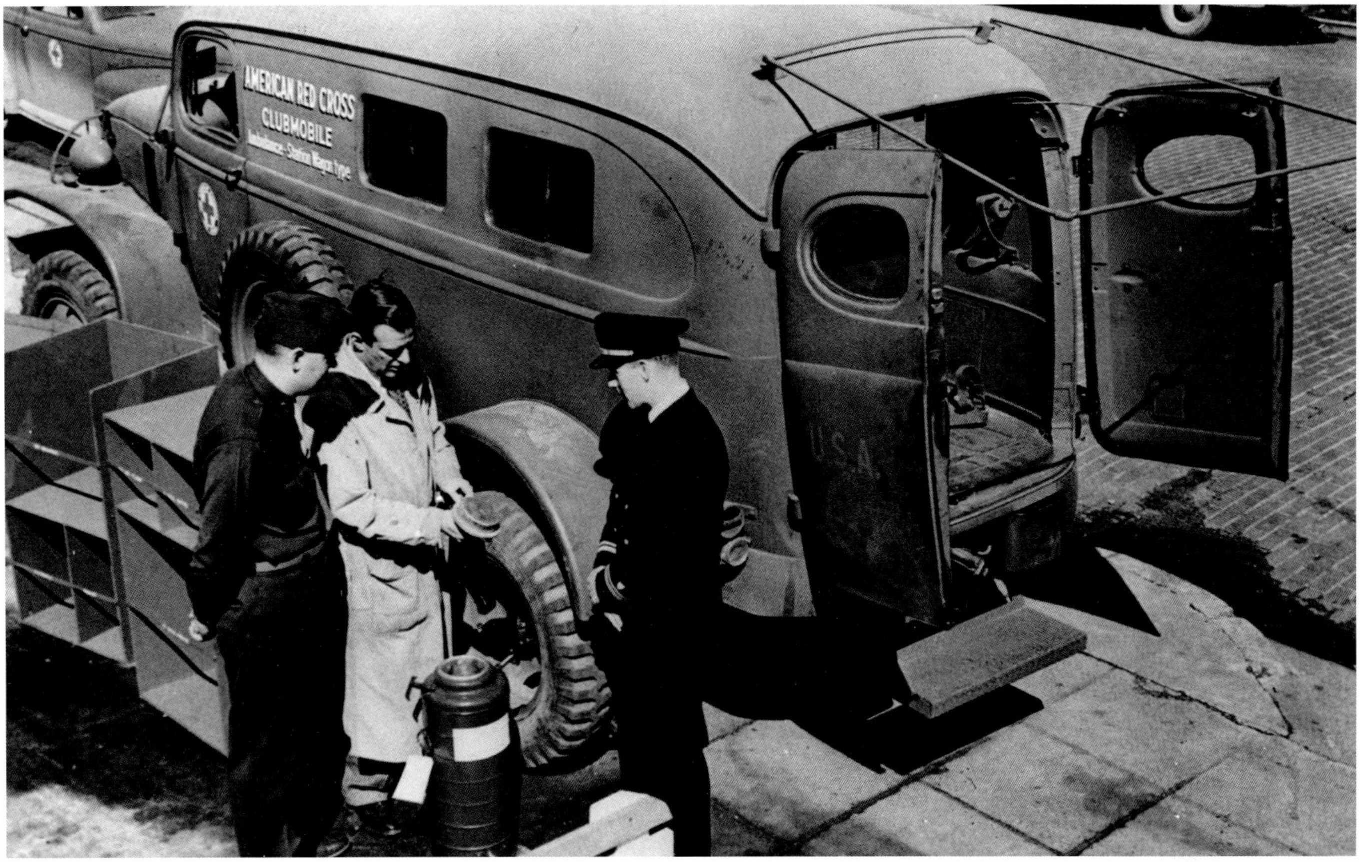

Above: *During World War II, the American Red Cross deployed quantities of Clubmobiles: military vehicles converted to mobile entertainment and refreshment facilities. This Chevrolet 1 1/2-ton 4x4 Panel Delivery Truck was converted to a Clubmobile and was destined for service along the Alaska Highway in September 1942. It was equipped with a donut machine, large coffee urns, a loudspeaker, and a movie projector. A roll-up movie screen could be set up inside the rear doors, and for showing movies during daytime, a sun screen could be rigged on the frame seen here on the rear of the roof. Painted on the body is "AMERICAN RED CROSS / CLUBMOBILE / AMBULANCE-STATION WAGON TYPE." In September 1942, the Red Cross contemplated sending at least six more Clubmobiles, in addition to this one, to the ALCAN Highway.* National Automotive History Collection

Below: *In a photo taken during a training session at the Basic Wire School at Camp Crowder, Missouri, on 14 October 1942, students are laying telephone wire using a Chevrolet 1 1/2-ton 4x4. Two reels of telephone cable are in the cargo body. One person is standing on the step on the rear of the vehicle. National Archives*

Above: *A master sergeant is laying telephone wire using a boom mounted on a Chevrolet 1 1/2-ton 4x4 truck at the Basic Wire School of the Signal Corps Replacement Training Center at the Camp Crowder Reservation in Missouri in October 1942. The boom is equipped with two pulleys for two lines. Note the wooden platform supported by chains on the rear of the vehicle. National Archives*

Below: *In a final view taken at Basic Wire School, Camp Crowder, Missouri, the same vehicle and personnel seen in the preceding photo are shown during a wire-laying exercise on 14 October 1942, with the boom mounted on the left side of the vehicle. The soldiers in the cargo body are identified as, left to right, Pvts. Mike Hidado, William Anderson, Edgar Dagle, and Darius McCormick. The "SCRTC" license plate refers to Signal Corps Replacement Training Center. National Archives*

Above: *During October 1942, a Chevrolet 1 1/2-ton 4x4 Cargo Truck, registration number 389066, is driving along a dirt road, towing a G-518 1-ton trailer. There are no visible markings on the truck other than the registration number, which corresponds to a vehicle produced under contract W-398-QM-11510.* National Archives

Below: *A Chevrolet 1 1/2-ton 4x4 Cargo Truck is parked across a dirt road during an exercise in October 1942. A registration number is present on the hood but is illegible.* National Archives

General Robert L. Eichelberger, commander of the U.S. I Corps, addresses his troops at Rockhampton, Queensland, Australia, on 19 October 1942. In the background is a Chevrolet 1 1/2-ton 4x4 Panel Truck. National Archives

Above: *A Chevrolet 1 1/2-ton 4x4 Cargo Truck proves its ability to extricate itself from a deep hole during the excavation of an emplacement for an antiaircraft gun at Fort Sheridan, Illinois, in or around October 1942. A placard with the number 6 on a white square, with two horizontal white bars to the right of the number, is attached to the upper part of the brush guard. The maker's mark, "CHEVROLET," stamped onto the side panel of the hood is clearly visible.* National Archives

Below: *This Chevrolet G-506 Cargo Truck, likely the same one in the preceding photograph, has just deposited a load of gravel in an excavation for an antiaircraft gun, at Fort Sheridan, Illinois. The tailgate is lowered, and gravity is providing the impetus for the unloading of the gravel.* National Archives

Above: *Members of an antiaircraft gun crew bearing rifles are being transported in a Chevrolet 1 1/2-ton 4x4 Cargo Truck with winch during practice firing at Fort Sheridan, Illinois, in or around October 1942. The registration number is 391716. A white placard bearing the numeral 4 is attached to the right side of the bumper with two bands.* National Archives

Below: *Scores of Chevrolet 1 1/2-ton 4x4 trucks are parked in the motor pool at Camp Blanding, Florida, in November 1942. Great variation is visible in the color tones and the stitching locations on the tarpaulins on these vehicles, reflecting different dye lots and sewing practices. The vehicles in the right foreground are equipped with winches.* National Archives

Above: *Personnel from the Signal Corps School at Fort Monmouth, New Jersey, are manning a mobile radio station in a Chevrolet 1 1/2-ton 4x4 Panel Truck, also known as the K-51, in a pine forest near Slagle, Louisiana, during Army maneuvers on 6 November 1942. Stenciled on the interior of the right rear door is information on the radio installation in the vehicle: an SCR-299-B set manufactured by Hallicrafters Company.* National Archives

Below: *A K-51 Panel Truck with a K-52 Trailer hitched to it is parked in a field at Fort Monmouth, New Jersey, around the summer or fall of 1942. A white marking, "X-2," is on the left bumperette. The communications-wire reels stored on the sides of the body were commonly painted in a darker color than the Olive Drab of the vehicles.* National Archives

The radio operator is at his station in the same K-51 shown in the preceding photograph. The staff sergeant who was handing a dispatch to the radio operator in the previous image is now standing on the K-52 Trailer. The side of the enclosure of the power plant is now open, providing a limited view inside. National Archives

Above: *Chevrolet 1 1/2-ton 4x4 Cargo Truck without winch, registration number W-351009, is poised behind what appears to be a smoke grenade during an exercise. This truck, a 1942 model, was produced under contract W-398-QM-10328.* National Archives

Below: *A Chevrolet 1 1/2-ton 4x4 Panel Truck set up as a radio vehicle and assigned to the 820th Signal Service Company is parked in Biskra, Algeria, during the Invasion of North Africa on 21 December 1942. Hitched to the truck is a 1-ton Trailer (G-518) with a spare tire stored underneath. The Signal Corps used these trailers to transport generators and other equipment. A whip antenna is on the roof of the panel truck, and a U.S. flag is superimposed over the identification star on the side door.* National Archives

Above: *A radio operator can be seen sitting cross-legged in the rear of a Chevrolet 1 1/2-ton Panel Truck fitted out as a radio vehicle for the Mobile Radio Unit of the 44th Signal Company, on detached service with the Twelfth Air Force. The vehicle and its 1-ton trailer are masked by camouflage netting and are parked in the courtyard of an art museum in a city in North Africa on 21 December 1942. Three whip antennas are mounted on the roof of the panel truck.* National Archives

Below: *This Chevrolet M6 Bomb Service Truck was damaged by shrapnel from a German bomb at the Youks-les-Bains Airfield in Algeria on 19 December 1942. The crew of the vehicle took cover under it during the air raid, escaping injury. The number 6889-A is stenciled on two places on the right side of the cowl; between those numbers is a dark-colored star with a number 3 on it.* National Archives

Above: Princess Evelyn *is the nickname painted on the driver's door of this Chevrolet 1 1/2-ton 4x4 Panel Truck in North Africa on 21 December 1942. This vehicle formerly was an Army Air Corps radio vehicle, used to coordinate close-air-support missions. After the truck was shot up during an enemy air attack, it was repaired, given a new engine, and converted to a mobile photographic lab.* National Archives

Below: *In a view of the opposite side of the same panel truck shown in the preceding photograph, "PRINCESS FLAVIA" is painted on the passenger's door. The door on the duffel compartment on the side of the body has three hinges on the rear side and three latches on the forward side, including a padlock hasp at the center.* National Archives

Telephone linemen from Company A, 56th Signal Battalion, are using a Chevrolet K-44 Pole-setter Truck with Earth Borer to prepare a hole for seating a telephone pole during a training exercise near Wylye, England, in late December 1942. The borer bit is well into the ground, and a pile of spoil is around the bit shaft. National Archives

Above: *U.S. Army mobile transmitters housed in Chevrolet 1 1/2-ton Panel Trucks are in operation at an undisclosed site in North Africa on 3 December 1942. Lockers called duffel compartments are present on the sides of the trucks; each of these has a door with three leaf hinges on the forward side and three latches on the rear side. A reel of communications wire is stored on toward the rear of each vehicle.* National Archives

Below: *A U.S. Army Signal Corps Chevrolet K-51 1 1/2-ton Panel Truck, carrying an SCR-299 radio transmitter-receiver, and a K-52 1-ton trailer with the generator and power unit for the radio, are parked in a field in England in January 1943. The registration number on the hood is 370709 S, the "S" indicating that the vehicle had been equipped and tested for radio-interference shielding.* National Archives

Above: *Japanese prisoners of war are being moved in several vehicles from a temporary camp to a permanent POW camp on New Caledonia on 7 January 1943. To the left is a Chevrolet 1 1/2-ton 4x4 cargo truck, W-340023, a long-wheel-base (175-inch) vehicle equipped with a front winch. The side panel below the hood has been removed from this vehicle. In the background is a Dodge 1/2-ton WC truck.* National Archives

Below: *U.S. Army truck convoy is hauling supplies from Fort McAndrew to a post exchange (PX) in Newfoundland on 12 January1943. The first three vehicles are Chevrolet 1 1/2-ton 4x4 cargo trucks. Large recognition stars are on the doors and cab roofs, while small stars are on the bumpers.* National Archives

Above: *In an aerial view of Youks-les-Bains Airfield in Algeria on 15 January 1943, a Chevrolet 1 1/2-ton Panel Truck equipped with radio transmitter-receiver equipment is parked in an area where aviation engineers are digging defensive shelters. Hitched to the panel truck is a 1-ton trailer with the generator for the radio.* National Archives

Below: *A Chevrolet M6 Bomb Service Truck tows a trailer loaded with aerial mines to a radar-equipped B-24 Liberator, "Tidewater Tillie," from the 2nd Anti-Submarine Squadron, 480th Anti-Submarine Group, at St. Eval, England, on 23 January 1943.* National Archives

Above: *The left rear wheel of this Chevrolet 1 1/2-ton 4x4 assigned to the19th Engineers was damaged by a land mine at Kasserine Pass, Tunisia, on 18 February 1943, the day before the beginning of the famous battle at that place. The wheel and the tire were wrecked. Only one occupant of the truck, a passenger, was injured. A 5-gallon liquid container and a storage chest are secured to the truck body. An American flag is on the driver's door.* National Archives

Below: *The same mine-damaged truck seen in the preceding photo is viewed from farther to the rear, showing the crater caused by the land mine. This explosion was a friendly-fire incident: the 19th Engineers had planted the mine shortly before the truck rolled over it, but the engineers had neglected to mark the mine.* National Archives

A K-51 Panel Truck containing an SCR-299 Radio Set, assigned to or working in conjunction with the 86th Bomb Squadron, is parked in an entrenchment for protective purposes, at Youks le Bains, Algeria, in February 1943. A camouflage net is rigged over the vehicle to mask it from crews of enemy aircraft. A small, stenciled marking, "6869-A," is stenciled on each side of the bumper. National Archives

Members of the 86th Bombardment Squadron are setting up the same entrenched K-51 seen in the preceding photo, for use as a communications station at Youks-les-Bains Airfield on 18 February 1943. One of the men is affixing the transmitting antenna onto its mount on the roof. National Archives

Above: *A convoy of G-506 Chevrolet trucks is transporting supplies from Oro Bay to Dobodura, New Guinea, on 14 February 1943. While the fourth truck in line has a standard steel cargo or dump body, the first three and the fifth have unusual, boxy bodies that appear to be of metal construction resting on visible cross members. The first truck has a pipe rack installed, and an unusual registration number on the hood: RC-1096 (although it is possible the first letter is H).* National Archives

Below: *This K-51 Panel Truck and K-52 Trailer with PE-95 Power Unit were part of the equipment assigned to the Amphibious Center at Camp Gordon Johnston, Florida, as photographed on 23 February 1943. All three mast antennas are secured with tie-downs.* National Archives

Above: *The reason why this Jeep is coupled by a tow bar to the pintle on the rear of K-51 registration number 3114846 at Camp Gordon Johnston, Florida, on February 14, 1943, is that the radio equipment in the panel truck was receiving electricity from a power unit stashed in the rear of the Jeep. This power unit is partially visible under the canvas top.* National Archives

Below: *Chevrolet K-51 registration number 3114846 is operating in sand dunes, and obviously struggling at it, at Camp Gordon Johnston with the same Jeep-mounted power unit shown in the preceding photo on February 14, 1943.* National Archives

Above: *Three U.S. Army Signal Corpsmen are rigging telephone lines along a railroad line in Tunisia on 13 March 1943. Parked on the far end of the buildings is a Chevrolet K-43 Telephone Maintenance Truck.* National Archives

Below: *A Chevrolet K-44 1 1/2-ton 4x4 Pole-setter Truck with Earth Borer and the "new" style body is in operation at Benghazi, Libya on 28 April 1943. The vehicle, including the derrick, likely was painted in a sand color. A USAAF-style recognition star is on the driver's door.* National Archives

The same K-44 shown in the preceding photo is viewed close-up as a 1st Sgt. Peak operates the earth borer. The registration number, 067480, and "USA" are painted on the rear of the body. The borer supports are deployed below the rear of the vehicle, to take up the load when the borer is excavating heavy earth. National Archives

Above: *During Third Army Maneuvers in Louisiana on 12 March 1943, a K-51 1 1/2-ton Panel Truck with a K-52 1-ton Trailer hitched to it are parked at Camp Polk, Louisiana. A pioneer tool rack is mounted vertically on the duffel compartment on the side of the panel truck.* National Archives

Below: *Private Jerome Lenzen is working on the engine of a Chevrolet Chevrolet 1 1/2-ton 4x4 Cargo Truck with Winch, registration number W-342924, at a Signal Corps motor pool in New Guinea on 4 May 1943. This vehicle was a G-7127 175-inch-wheelbase model, as indicated by the registration number. Next to his left hand is the AC oil filter. Details of the hold-open arm for the hood are in view.* National Archives

Private 1st Class Garnet Shatzer is operating a K-44 earth borer during Second Army maneuvers in Tennessee on 26 May 1943. He is holding the two levers that controlled the borer. For example, the auger shaft was lowered by pulling up the left-hand lever and pushing down the right lever. Pushing down both levers both lowered the auger and set the auger to rotating, in order to bore into earth. National Archives

Above: *A Chevrolet K-51 1 1/2-ton Panel Truck with an SCR-299 radio transmitter-receiver and a K-52 1-ton trailer with generator, serving with the British 4th Light Armoured Brigade, is set up in a field north of Djembinina, Tunisia, on 9 May 1943. The crew of the vehicle was maintaining communications with corps headquarters.* National Archives

Below: *On 12 May 1943, a Chevrolet 1 1/2-ton dump truck is unloading crushed rock on a 20-ton, all-weather Army road near Oro Bay, New Guinea. The construction of this road was a project of the 96th Engineers, of the U.S. 41st Division.* National Archives

A member of Company A (Construction and Lineman Company), 63rd Signal Battalion, demonstrates the operation of a K-44 earth borer in sandy soil at Oujda, Morocco, on 18 June 1943. When boring holes in sandy soil, the auger was typically set to rotate at about 125 rpm. National Archives

In a companion photograph to the preceding one, a Telephone Maintenance Truck K-43 is on display at Oujda, Morocco, on 18 June 1943. The boom has been erected, and the doors on the storage compartments on the side of the body are open. Above the "30 MPH" stencil on the rear of the body is a small placard, apparently installed for this exhibition, that reads, "CONSTRUCTION TRUCK." National Archives

Above: *The post engineer at the Ordnance Depot, Little Rock, Arkansas, designed the modification seen on this Chevrolet Class 110 Airfield Crash Truck, registration number 501317, photographed at Adams Field, Little Rock, on 11 July 1943: a foam-chemical tank on top of the body. This tank was above the main water tank and was connected by a valve to the water tank, allowing the foam chemical to flow into the water tank. The foam-water solution, if allowed to sit, congealed. The purpose of this rig was to pour the foam into the water at the beginning of an emergency response, to make sure the solution was fresh and fluid.* National Archives

Below: *The same modified Chevrolet Class 110 Airfield Crash Truck shown in the preceding photograph is viewed from the rear, showing how the foam tank is mounted above the body. A close examination of the photo reveals that legs in the form of metal bars are bolted to the rear corners of the foam tank to support it. The hand wheel above the hose reel, not present on stock Class 110 trucks, evidently controlled the valve for releasing the contents of the foam tank into the main water tank.* National Archives

Above: *As the strategic-bombing campaign against Germany ramped up during 1943, Chevrolet M6 Bomb Service Trucks played a role, transporting bombs to the bombers at airfields across England. On this example, registration number 0012171S, an ordnance man is operating the hoist to unload a 500-pound bomb from the trailer, while arming a B-17 Flying Fortress of the 381st Bomb Group on 20 August 1943.* National Archives

Below: *Although, as we saw earlier, Aberdeen Proving Ground experimented with an Auto-Railer version of the Chevrolet 1 1/2-ton 4x4 in early 1945, this photo is evidence that these trucks were being used on railroads more than a year earlier. This Chevrolet Panel Delivery Truck, converted for rail use, was being used by the Alaska Communication System (ACS) in September 1943 to repair and maintain its telegraph lines along the Alaska Railroad. The vehicle carried registration number W-348827, and official markings are on the driver's door.* National Archives

Around October 1943, Signal Corpsmen are laying communications wire along a road in a forest in an unidentified area during maneuvers. The vehicle is a Chevrolet 1 1/2-ton 4x4 Cargo or Dump Truck with a RL-26 reeling machine for two reels of communications wire in the cargo body. National Archives

Above: *At a rock crusher operated by U.S. Army Engineers in New Guinea in December 1943, a Chevrolet 1 1/2-ton 4x4 Dump Truck, left, is delivering a load of stones. A close examination of the photograph reveals that chains are installed on the tires, to improve traction in soft and muddy soil. Another dump truck is underneath the crusher to the far right.* National Archives

Below: *At a battery of rock crushers along the Ledo Road between India and Burma sometime during 1943, a Chevrolet 1 1/2-ton 4x4 Dump Truck is positioned underneath a chute below one of the crushers.* National Archives

Above: *Members of an African-American unit are carrying rocks from a hillside to a portable rock crusher that has been loaded on a 6-ton 6x6 truck. The crushed rocks then are being loaded on the Chevrolet 1 1/2-ton 4x4 dump truck that is backed up to the 6-ton truck. Chains have been mounted on the tires of the two trucks, for better traction on the muddy road.* National Archives

Below: *A lineman is inspecting telephone lines strung along the rear wall of the Palazzo Reale in Caserta, Italy, in January 1944. Parked alongside the palace is a Chevrolet K-43 with high-visibility recognition stars in circles on the door and the hood and two five-gallon liquid containers, one above the other, on the front of the body. Markings on the bumper are "5A 515," two white horizontal bars, a recognition star, and "B 4."* National Archives

Signal Corps linemen are installing telephone lines at Slapton Sands, England, in January 1944, using a Chevrolet K-43. Stencils indicating a maximum speed of 30 miles per hour, and that this is a left-hand-drive vehicle, are on the rear of the body. The British built a base at Slapton Sands in late 1943, for the use of the U.S. forces that would be landing on Utah Beach in Normandy, in 1944. National Archives

Telephone linemen from Company A, 96th Signal Battalion, are employing the boom on the rear of a Chevrolet K-43 to run telephone lines through a brushy area at MP 9 along the Ledo Road in February 1944. National Archives

Above: *An armorer is sitting on one of two 2,000-pound general-purpose bombs on an M5 Bomb Trailer being towed by an M6 Bomb Service Truck at a base in England in the summer of 1943. An M1 Bomb Lift Truck is stowed behind the bombs on the trailer. In the background is a B-17F "Quitchurbitchin", of the 91st Bomb Group.* National Archives

Below: *Staff Sgt. Thamar White, from the 832nd Signal Service Detachment, stands by, prepared to operate the earth borer on a late-production Chevrolet K-43 at Townsville, Australia, on 30 March 1944. On the boring case is painted a light-colored triangle with the number 2 on it. Under the right rear of the body, the pole-derrick supports are collapsed and secured with chains.* National Archives

Above: *A Chevrolet M6 Bomb Service Truck is parked next to "Chennault's Pappy," a B-17 Flying Fortress from the 91st Bomb Group, at Bassingbourne, England, on 9 April 1944. Hitched to the truck are two M5 Bomb Trailers and, to the rear, a tricycle bomb dolly. A variety of types of bombs are on the trailers and dolly. The truck is marked for the 30th vehicle of the 401st Bomb Squadron, 91st Bomb Group.* National Archives

Right: *Armorers from the 450th Bomb Squadron, 322nd Bomb Group, are about to load bombs from an M5 Trailer into a Martin B-26 Marauder at Andrews Field, Essex, England, during March 1944. The trailer has been unhitched from the M6 Bomb Service Truck in the background, and the men are pushing it into position next to the aircraft. The name "FRENCHIE" is painted in capital letters on the rear of the platform of the trailer.* Roger Freeman collection

Above: *Ordnance men from the 91st Bomb Group sit on bombs on an M5 Bomb Trailer hitched to a Chevrolet M6 Bomb Service Truck as a B-17 Flying Fortress takes off from Bassingbourne, England, on 9 April 1944.* National Archives

Below: *Ordnance men are loading 500-pound bombs onto M5 Bomb Trailers coupled to an M6 Bomb Service Truck on Saipan, in the Mariana Islands, on 5 July 1944. These bombs were to be loaded onto P-47 Thunderbolts for a mission to nearby Tinian.* National Archives

Above: *A radio operator is visible in the rear of a Chevrolet K-51 Panel Truck in the jungles of New Georgia, Solomon Islands, in the Southwest Pacific during June 1944. Hitched to the truck is a K-52 Trailer with a PE-95 Power Unit. Several five-gallon liquid containers are sharing space with the power unit in the trailer. Rolled-up tarpaulins from the trailer are stacked to the far right.* General Motors LLC

Below: *Technician 4th Class Alfred Rizzio of the 96th Signal Battalion is operating the earth borer on a late-model Chevrolet K-44 along the Ledo Road in Burma while, from left to right on the ground, Lt. Col. George F. Moynahan, chief signal officer of the Northern Combat Area Command; Staff Sgt. Wilbur T. Hansen; and Pfc. Fred B. Walker, look on. The date was 23 July 1944. Visible in the left side of the truck body is the power reel CR, for quickly paying-out or taking-up telephone wire.* National Archives

Above: *The booms of two K-43 Telephone Maintenance Trucks are being used to load a portable communications office at Oudja, Morocco, on 26 July 194. The photograph was taken by a cameraman from the 163rd Signal Photo Company, so it is likely these trucks were assigned to the 163rd Signal Battalion.* National Archives

Below: *Members of a telephone-line salvage crew from Fort Lewis, Washington, are rewinding telephone wire during an exercise at Camp Adair, Oregon, on 2 August 1944. On the right side of the K-43 truck body, affixed to the winch shaft, is the power reel CR; "CR" stands for collapsible reel.* National Archives

Technician 5th Class W. J. Shott of the 432nd Signal Construction Battalion is at the controls of an earth borer on a late-production Chevrolet K-44 outside of Calcutta, India, on 15 August 1944. Although the truck has sunk into mud and is out of level, it was possible to cross-level the borer in order to set the telephone poles in perfectly plumb position. National Archives

Above: *A late-type Chevrolet K-44 is being hoisted from the hold of the Liberty Ship* Henry T. Scott *at Calcutta, India, in August 1944. This was one of four U.S. Army Signal Corps trucks brought into the port aboard this ship. The pole derrick is resting on a travel lock on the top front of the body. At the front of the body is the winch housing. On the left side of the body are the collapsible power reel and the stored augers. To the rear of the spare tire is a beverage container, secured to a stand. On the roof of the cab is the Cab-O-Scope, a rear-view mirror assembly that allowed the driver to see over the rear of the vehicle and to discern hand signals from the earth-borer operator.* National Archives

Below: *Among the U.S. Army vehicles arrayed along the beach at Place de Napoléon in Cherbourg, France, on 13 August 1944, is a G-506 Chevrolet 1 1/2-ton 4x4 Truck-Tractor, just beyond and to the left of the crane in the foreground. Hitched to this truck-tractor is a stake semi-trailer. To the immediate front and rear of the G-506 are AFKWX 2 1/2-ton trucks. Closer to the beach are numerous DUKWs.* National Archives

Above: *Radio operator Tech 5 Irving Katsper is at work in the rear of a Chevrolet K-51 1 1/2-ton 4x4 Panel Truck in the Caserta area of Italy on 18 August 1944. The radio equipment installed in the vehicle, as specified on the stencil on the right rear door, is the SCR-299-C. The stencil also includes the name of the radio's manufacturer, Hallicrafters, and the order numbers of the radio equipment.* National Archives

Below: *A column of 1 ½-ton 4x4 Chevrolet tractors and trailers, part of the Red Ball Express, are refueled for another trip in France in 4 September 1944. From the stenciling on the windshields of these tractors, the intended cargo is ammunition. The Red Ball Express utilized the Chevrolet in Light Truck Companies. The Red Ball Express operated from August through November 1944.* National Archives

Above: *Ordnance men from the Seventh Air Force are unloading 500-pound bombs from an M5 Bomb Trailer hitched to a Chevrolet M6 Bomb Service Truck, for transfer to the Consolidated B-24 Liberator in the background, at an airfield in the Marianas in September 1944. A spotlight is attached to the underside of the I-beam hoist support, for use in nighttime operations such as this one.* National Archives

Below: *Chevrolet Class 110 Airfield Crash Truck registration number 501414 is providing fire-extinguishing foam for an emergency crew striving to squelch flames after a B-17G, serial number 44-6316, of the 358th Bomb Squadron, 303rd Bomb Group, has crashed due to landing-gear failure during landing, at Molesworth, England, on 12 September 1944.* National Archives

A Chevrolet M6 Bomb Service Truck and M5 Bomb Trailer have brought a load of 100-pound general-purpose bombs to a B-17 Flying Fortress nicknamed "Chug-a-Lug IV," from the 535th Bomb Squadron, 381st Bomb Group, at Ridgewell, Essex, England, in October 1944. This Flying Fortress would be lost over Europe in November 1944. "535th ORD" is stenciled in white in two places on the rear of the bomb trailer. National Archives

The same bomb-unloading operation depicted in the preceding photo is viewed from a slightly different angle. A spotlight is on the right side of the I-beam of the bomb service truck. "535 SQ." and the number 13 are stenciled on the box to the rear of the front seats. National Archives

Above: *At a firing range at Laredo Army Air Force Base, a Chevrolet E-5 with a canvas cover over the frame for the turret is parked in the left background, behind a covered gallery in which six numbered, static machine-gun turrets are positioned.* National Archives

Below: *Ordnance men at a bomb dump on Saipan are lifting bombs onto a trailer using the hoist on a M6 Bomb Service Truck in November 1944. The truck was assigned to the 873rd Bomb Squadron, and these bombs were destined for B-29 Superfortresses based at Iseley Field, Saipan. "873-38" is marked on the hood and on the bumper.* National Archives

Above: *An ordnance man cranks the handle of the hoist on a Chevrolet M6 Bomb Service Truck at a bomb dump on Angaur, in the Palau Islands, in November 1944. This truck served with the 864th Bomb Squadron, as indicated by the number 864 on the rear of the truck.* National Archives

Below: *The same man and M6 Bomb Service Truck seen in the preceding photo are to the front of the row of bombs to the left of this photo, on Angaur in November 1944. To the far left is an M5 Bomb Trailer, and in the left background is another Chevrolet Bomb Service Truck.* National Archives

Telephone earth-borer trucks were not used strictly for emplacing telephone poles. Members of the 120th Engineer Battalion are using this Chevrolet K-44 1 1/2-ton Telephone Earth Borer to place holes in a road to be filled with explosive charges, to be detonated if U.S. troops are forced to retreat from the position. The site was in the Hirschthal area of Germany, on 30 December 1944. National Archives

Above: *At a bomb dump on Angaur on 2 December 1944, ordnance men from the 864th Bomb Squadron, 494th Bomb Group, are loading 1,000-pound bombs onto an M5 Bomb Trailer hitched to an M6 Bomb Service Truck. The bombs were to be loaded onto B-24 Liberators for airstrikes against Japanese forces on Babelthaup, in the Palau Islands.* National Archives

Below: *At the same bomb dump on Angaur shown in the preceding photo, members of the 864th Bomb Squadron roll a 1,000-pound bomb onto an M5 Bomb Trailer on 2 December 1944. The caster wheel on the trailer is turned to the side. To the right is an M6 Bomb Service Truck marked "864-45" on the front bumper, with a flag mounted on the bumper.* National Archives

Above: *In a posed photograph, possibly taken during stateside maneuvers, an infantryman points the way forward to the driver of a Chevrolet K-51 Panel Truck towing a K-52 Trailer with PE-95 Power Unit. The K-51 provided the Army with a fast-moving mobile radio communications platform, with a range capability exceeding 100 miles.* US Army Quartermaster Museum

Below: *A K-51 Panel Truck and K-52 Trailer combination is parked in the jungle somewhere in the Southwest Pacific in December 1944. Palm fronds are arranged on the truck and the trailer for camouflage purposes. "60 SIG" is chalked on the side and the rear of the trailer, likely a reference to the 60th Signal Battalion, which was active in that theater of operations.* US Army Quartermaster Museum

Above: *At an air base in Belgium in late 1944, a P-38 Lightning fighter plane has crashed into several trucks, including two Chevrolet M6 Bomb Service Trucks in the foreground. A placard that reads "EXPLOSIVES" is on the grille of the truck at the center foreground. In the center background is a G-506 Chevrolet 1 1/2-ton 4x4 truck equipped as a crash truck or emergency vehicle: a hose is connected to a fitting in front of its grille. Also in the scene are three CCKW-based decontamination trucks, which apparently have been re-purposed into fire service.* National Archives

Below: *Ground crewmen of the 55th Fighter Squadron, 20th Fighter Group, pose next to an M6 Bomb Service Truck, registration number 0036750, during 1944. This squadron was based at Kings Cliffe, Northamptonshire, England, from the summer of 1943 to the end of World War II.*

Above: *A Chevrolet Class 135 Crash Truck is parked near the control tower at RAF Little Walden, home of the 361st Fighter Group, in Essex, England, during 1944. The registration number on the hood is blurred, and the next-to-last number is indistinct, but it appears to be 5083X9. As a modification, a protective frame has been built around the pump on the front end.*

Below: *Ground crewmen of the 391st Bomb Group, Ninth Air Force, are loading .50-caliber ammunition into boxes ammunition boxes in 1944. Some of the loaded boxes have been placed on the M5 Trailer in the foreground. On the far side of that trailer, An M6 Bomb Service Truck is towing an M5 Trailer full of bombs. Another M6 is in the right background. In the background is a small sign that reads, "TAXI SLOWLY."*

Above: *Two signal men from Company C, 2nd Signal Battalion, U.S. Ninth Army, are boring a hole with a Chevrolet K-44 1 1/2-ton 4x4 Pole-setter Truck with Earth Borer outside of Dorff, Germany, on 3 January 1945. A shipping stencil is visible on the forward part of the body. The "XIX 2S" marking on the rear of the vehicle refers to XIX Corps, 2nd Signal Battalion.* National Archives

Below: *Armorers from the 468th Bomb Group are unloading bombs from an M5 Bomb Trailer behind an M6 Bomb Service Truck at an airbase in India on 6 February 1945. Armorers would load these bombs on racks on B-29 Superfortresses, which on the following day were scheduled to bomb Bangkok, Siam (now Thailand).* National Archives

Above: *A Chevrolet M6 Bomb Service Truck with an M5 Trailer hitched behind it is bringing 500-pound bombs to a flight line of P-47 Thunderbolts of the 1st Air Commando Group, Tenth Air Force, at an airbase in India on 10 February 1945. The registration number 0062024 is stenciled on the hood of the truck. The first Thunderbolt bears the nickname* Yankee Witch. Roger Freeman collection

Below: *A Chevrolet K-43 Telephone Maintenance and Construction Truck serving with the 443rd Signal Heavy Construction Battalion on Tinian in the Mariana Islands is hitched to cable dolly in early February 1945. Visible inside the truck body is the spare tire and coils of rope. Mounted on the lower right rear of the body is a vise.* National Archives

Above: *Chevrolet K-43 registration number 0033539 of the 443rd Signal Heavy Construction Battalion is viewed from the left side with the storage compartment doors open, on Tinian in February 1945. The presence of the extension ladder, derrick sections, and liquid containers on the running board made it impossible for the driver to use the left door. Stenciled in white above the rear fender (and also on the rear and the right side of the body) is "LIRP 94959."* National Archives

Below: *On Tinian in early February 1945, a Chevrolet K-43 is hauling telephone poles on a trailer. Four GIs are riding in the body of the truck.* National Archives

Above: *Visible inside the rear of this K-43 of the 443rd Signal Heavy Construction Battalion on Tinian are the winch, a collapsible reel on the left side of the compartment, and many coils of rope. A number of shovels are stored in the cubbyhole in the upper right of the body.* National Archives

Below: *The doors on the right side of the body of the same K-43 are open, revealing the storage compartments, bins, and parts drawers within. A drinking-water container is inside the rear door. On a rack on the upper forward door is a first-aid kit.* National Archives

Above: *Members of the 3rd Construction Section, 443rd Signal Heavy Construction Battalion, pose next to Chevrolet K-43 registration number 0033539 on Tinian in early February 1945. Some of the men are wearing lineman's belts and climbers.* National Archives

Below: *A Chevrolet K-44 Pole-setter Truck with Earth Borer assigned to the 443rd Signal Heavy Construction Battalion on Tinian is viewed from above during February 1945. Like the K-43 featured in the preceding photos, this K-44 had the code "LIRP 94959" stenciled in white on the side of the body: in this case, above the doors of the forward storage compartment. It is not clear what this marking signified. "US ARMY" was stenciled in white on the cab door.* National Archives

Above: *Members of the 443rd Signal Heavy Construction Battalion pose next to their Chevrolet K-44. The derrick has been raised to its operating position. One member is standing at the control levers of the earth borer. National Archives*

Below: *Members of the 1st Platoon, Company B, 432nd Signal Heavy Construction Battalion, are employing the boom on a K-43 Telephone Maintenance Truck with winch to load telephone poles on a bolster truck with trailer on the Burma Road near Lungling, China, on 16 March 1945. Note the chocks snugged up to the rear tires of the K-43. National Archives*

Above: *Chinese children watch the same telephone-pole-loading operation depicted in the preceding photograph, near Lunging, China, on 16 March 1945. The men are in the process of hoisting an extra-long telephone pole to a trailer. National Archives*

Below: *Members of the 2nd Platoon, Company B, 432nd Signal Heavy Construction Battalion, stand in line for chow, served from the truck to the left, along the Burma Road on 10 March 1945. To the far right is a K-43 Telephone Maintenance Truck with the boom erected, to the rear of which is a Dodge 3/4-ton truck. National Archives*

Private 1st Class Leonard M. Fowler, of the wire-stringing crew of the 2nd Platoon, Company B, 432nd Signal Heavy Construction Battalion, supervises the operation of the power reel CR on the right side of a Chevrolet K-44 on the Burma Road on 15 March 1945. Connected to the winch shaft, the power reel CR enabled the crew to dispense or take-up telephone wire quickly. National Archives

Above: *At a site in France on 8 March 1945, men of Company C, 57th Signal Battalion, VI Corps, Seventh Army, are engaged in a training exercise with captured German communications equipment. The man in the foreground is signaling to the operator of the earth borer on the K-44. Stenciled on the borer case is "VI575."* National Archives

Below: *The same earth borer seen in the preceding photo has completed a hole, and the derrick and winch are being used to lower a telephone pole into it. In the foreground, men are using pike poles and a support to manage the top-heavy pole as it is maneuvered into position.* National Archives

Above: *A military policeman watches as Signal Corpsmen unload telephone poles from a vehicle at Chabua, Assam, India, on 30 March 1945. A K-43 with a boom is performing the lifting operations.* National Archives

Below: *In March 1945, U.S. troops at Rosario, Luzon, liberated this Chevrolet G-506 truck that the Japanese had captured earlier in the war and had put to use during their occupation of the Philippine Islands. The vehicle had seen hard use: the grille is battered, the right side panel of the hood is missing, and the windshield is smashed. The faint circle on the cab door was likely a red "meatball" insignia.* National Archives

Corporal Roy Field stands on the left running board of M6 Bomb Service Truck registration number 0036802 at Fritzelar, Germany, in April 1945. This vehicle was accepted during 1943 as part of contract W-374-ORD-4052. National Museum of the United States

Above: *The M6 Bomb Service Truck was capable of towing two fully loaded M5 Bomb Trailers, hitched in tandem, as seen in this photo taken in April 1945. Each of the trailers is loaded with eight 500-pound general-purpose bombs, with the fins stored on their holders on the storage boxes at the front of the trailer beds.* National Museum of the United States

Below: *Firefighting equipment is being delivered on a barge in the Port of Livorno, Italy, on 2 May 1945. The vehicles to the far left and the far right are Class 135 Crash Trucks, and the two in the center are class 325 Fire Trucks with extension ladders stored on the right sides.* National Archives

Above: *In a view looking to the southwest from the control tower at Kadena Airfield on Okinawa on 9 May 1945, in the left foreground is a Chevrolet Class 135 Crash Truck. "BIVE" is roughly painted on the two doors of the cab. In the right background are a C-2 wrecker, two crash or fire trucks, a crane, and a CCKW cargo truck. In the distance is a row of F4U Corsair fighter planes.* National Archives

Below: *A Chevrolet K-43 Telephone Maintenance Truck, registration number 0033548, is parked in the foreground of this view of members of Company A, 96th Signal Battalion, replacing a massive H-fixture at Myitkina, Burma, on 7 June 1945. The H-fixture would support telephone lines spanning the Irrawaddy River, facilitating communications between China and Myitkina. This line would span 2,300 feet.* National Archives

Above: *A Chevrolet K-51 1 1/2-ton 4x4 Panel Truck is parked in a communications area on Kume Shima, part of the Okinawa Islands, on 29 July 1945. To the front of this truck in the distance is another K-51. A reel for communications wire is on the right rear of the body, and a wooden ladder is propped up against the rear of the vehicle for the convenience of the radio operator. Numerous radio antennas are installed on the ground.* National Archives

Below: *After the conclusion of the war in Europe, much of the war materiel used in that theater of war was processed for shipping to the Pacific Theater of Operations, for the anticipated invasion of Japan. Here, U.S. Army personnel at the Port of Marseilles, France, are preparing a Chevrolet K-43 Telephone Maintenance Truck for shipment to the PTO on 18 July 1945. The registration number, 0097869 S, pertains to a 1944-model K-43.* National Archives

In the first of a series of three photographs documenting the work of the 463rd Signal Construction Battalion in the Obascom area of Okinawa on 27 November 1945, a crew is rigging a cable to telephone poles using a large reel on a trailer. National Archives

Linemen from the 463rd Signal Construction Battalion are rigging cables atop a telephone pole in the Obascom area of Okinawa on 27 November 1945. In the foreground is the rear of a K-43 Telephone Maintenance Truck, with an extension ladder stored on the hooks on the side of the body and a large coil of heavy rope stowed on the rear. National Archives

Above: *A signalman in the passenger's seat of a 1944-model Chevrolet K-43 Telephone Maintenance Truck holds onto a wire being unspooled from the power reel CR on the side of the truck body. "U.S. ARMY" is painted in white on the door. The registration number on the hood is obscure, but it appears to be 0080414. Faintly visible to the upper right of the power reel is stenciled "463 S," signifying the 463rd Signal Construction Battalion.* National Archives

Below: *At a U.S. Army depot in New Delhi, India, in 1945, Lt. A. R. Plagar and Maj. C. E. Dugan inspect Blaw-Knox antenna sections declared surplus and ready for liquidation. In the background is a K-43 Telephone Maintenance Truck, registration number 0098160 S.* National Archives

Above: *Members of the 2794th Engineer Fire-fighting Platoon pose on what appears to be a Chevrolet Class 325 4x4 Fire Truck at Heidelberg, Germany, on 21 January 1946. This platoon had seen combat in the Ardennes-Alsace Campaign in the final months of World War II.* National Archives

Below: *Members of a fire-fighting unit from the 1207th Engineers pose next to their vehicles in front of a fire station in Livorno, Italy, on 27 February 1946. To the far left is a tanker truck, to the right of which is a Chevrolet 1 1/2-ton 4x4 fire truck, a Ford Class 325 4x2 Fire Truck, and a jeep.* National Archives

Above: *A Chevrolet K-51 Panel Truck towing a K-52 Trailer for a PE-95 Power Unit and carrying signs extolling the U.S. Army Signal Corps is part of a 20-vehicle Signal Corps contingent in the Army Day Parade in Sacramento, California, on April 6, 1946.* National Archives

Below: *This and the following two photos, taken around July 1946, document the installation of enclosures on the roofs of some Chevrolet K-51 Panel Trucks. The photos were part of a display of photos at the Signal Corps Depot in Philadelphia at that time, so it is presumed that the images were taken at that facility. The enclosures had slightly curved roofs and hinged side panels. There were two of these panels per side, each one having three leaf hinges at the top. In the right foreground is a 1942 K-51, registration number 370393.* National Archives

Above: *K-51 370393 is seen from the right side in the foreground. Marked in chalk on the side of the passenger's seat is "Greased" over "REAR + TRANS." More K-51s are being fitted with top enclosures in the background. A close examination of these photos reveals that at least the side panels of the enclosures were made of plywood. In addition, there was a box-like structure on top of the running board from the rear of the cab door to the rear fender.* National Archives

Below: *More K-51s being fitted with top enclosures are depicted. As seen on the closest vehicle, registration number 370623, the front panel and roof (and, out of view, likely a rear panel) were attached to the panel truck roof before the hinged side panels were installed. The roofs of the panel trucks were not cut away, as would be the case if these enclosures were for providing more headroom for the radio compartment, so likely the enclosures were for storage purposes.* National Archives

Above: *Two sailors pose in an undated photograph next to a K-51 Panel Truck with markings on the door for the U.S. Naval Reserve's Organized Surface Battalion stationed in Dallas, Texas. To the rear of the vehicle is the left wheel of a trailer that probably is carrying the power unit for the K-51's radio equipment.* National Archives

Below: *This is the first in a series of five photos documenting a 1944 Chevrolet K-44 Pole-setter Truck with Earth Borer, registration number 0099047, modified with a work platform on the derrick to facilitate the work of linemen. In this view, part of the platform is visible above the forward part of the body. Photographed in Seoul, Korea, on 2 September 1948, the vehicle displays signs of hard service: a crumpled driver's door and dented bodywork.* National Archives

The 1944 Chevrolet K-44 with the work platform on the derrick is viewed from the rear in the lowered position at Seoul on 2 September 1948. The platform was fabricated from wood. Unit markings on the rear of the truck body are "XXIV" and "B66." The platform was developed by Sgt. Byron G. Anderson of the 76th Signal Service Battalion. National Archives

Two linemen are demonstrating the operation of the work platform on the derrick of a K-44. The platform was attached to a vertical beam, which, in turn, was fitted with U-type metal straps to loosely secure it to the derrick. A winch cable was routed over the sheave at the top of the derrick and hooked to the vertical beam, and by this simple mechanism the platform could be raised and lowered on the derrick. National Archives

In this view, the winch cable has raised the work platform to a position from which the lineman can inspect some lead-covered cables.
National Archives

In the final photo of the sequence, the rig operator has raised the work platform some more and has tilted the derrick and the work platform to enable the lineman to easily reach a cable. National Archives

A telephone construction crew is inspecting and tightening lead cables by hand at a communications center in Korea on 6 August 1950. In the foreground, an operator has his hands on the control levers of the earth borer on a Chevrolet K-44. National Archives

In a photograph related to the preceding one, a lineman on a bamboo ladder is installing lead cable; at the bottom of the photo is the top of the cab of a Chevrolet K-44. The windshield is secured in the open position. National Archives

Above: *In a photograph of South Korean troops receiving instruction on 15 September 1950, a Chevrolet 1 1/2-ton 4x4 Truck is parked in the left background. A very irregularly applied recognition star is painted on the passenger's door.* National Archives

Below: *Signalmen are constructing a guard around a transmitter tower in Korea on August 20, 1950. They will be lowering wooden poles into the holes around the tower, which presumably were bored by the Chevrolet Polesetter Truck with Earth Borer K-44. The bumper of the vehicle is marked "71S," for the 71st Signal Battalion.* National Archives

The Chevrolet K-44 on the left, registration number 0099438, has been put to novel use, its derrick serving as a mast for a parabolic antenna during microwave field tests in February 1953. On the service body of the truck to the right is receiver equipment. National Archives

Above: *A Chevrolet Class 135 Crash Truck stands by as aircraft burn following the crash of a F-51 Mustang fighter plane from the South African Air Force at Seoul City Air Base in September 1950.* National Archives

Below: *By the end of the Korean War, newer and more sophisticated military vehicles were supplanting the Chevrolet G-506 trucks, which were over a decade old. The 1941 Chevrolet 1 1/2-ton 4x4 Dump Trucks in this photo were awaiting public auction at the Ogden Arsenal in Utah on 2 December 1954. The closest vehicle bears USAF registration number B3027.* National Archives

Tables

Table 1 - CHEVROLET G-506 GENERAL DATA

MODEL	145" wb cargo	175" wb cargo	175" wb COE	125" wb M6
WEIGHT NET	7545	8150	8570	6325
MAX TOWED LOAD	8000	8000	8000	8000
LENGTH*	224	296	286	221
WIDTH*	86	86	96	75.75
HEIGHT*	106	106	99	91.5
WIDTH#	49.25/85.75	49.25/85.75	49.25/85.75	49.25/85.75
TRACK	67.5	67.5	67.5	67.5
TIRE SIZE	7.50 - 20	7.50 - 20	7.50 - 20	7.50 - 20
MAX SPEED	48	48	43	50
FUEL CAPY	30	30	30	30
RANGE	270	270	240	240
ELECTRICAL	6v Neg. Ground	6v Neg. Ground	6v Neg. Ground	6v Neg. Ground
TRANSMISSION SPEEDS	4	4	4	4
TRANSFER SPEEDS	2	2	2	2
TURNING RADIUS FEET	31	36.5	36.5	27.5

- *Overall dimensions listed in inches.
- # inside/outside width at tires.

ENGINE DATA

ENGINE MAKE/MODEL	Chevrolet BV
NUMBER OF CYLINDERS	6
CUBIC INCH DISPLACEMENT	235.5
HORSEPOWER	83 @ 3100
TORQUE	184 @ 1000
GOVERNED SPEED (rpm)	3100

Table 2 - G-506 production distribution by contract and plant **Continues →**

Chevrolet Plant Code Number	8	14			1
Location	**Atlanta**	**Baltimore**	**Cleveland**	**Detroit ***	**Flint**
DA-W-398-QM-8				x	
DA-W-398-QM-243					
DA-W-398-QM-421					
W-374-ORD-6582	x				
DA-W-398-QM-517	x	x			x
W-374-ORD-1131					x
W-374-ORD-1185		x			
W-374-ORD-1281		x			
W-374-ORD-2216					
W-374-ORD-2694	x	x			
W-374-ORD-2694S	x	x			
W-374-ORD-2744		x			
later ORD-2744 W-398-QM-11508			x		
W-374-ORD-2771		x			
W-374-ORD-2797					
W-374-ORD-2816					
W-374-ORD-2831					
W-374-ORD-2839					
W-374-ORD-2973					x
W-374-ORD-4052					x
M6 parts W-374-ORD-5146					x
W-374-ORD-5151				x	
W-374-ORD-6582	x				
W-398-QM-8278				x	
W-398-QM-8686				x	
W-398-QM-8778				x	
W-398-QM-9402				x	
W-398-QM-10328				x	
W-398-QM-10708		x			
W-398-QM-10939					x
W-398-QM-11546					
Nos-81721				x	
Nos-95703		x			
Nos-95878	x	x			
Nos-2919		x			

* No actual production of trucks occurred in Detroit, see the Introduction to Table 4 for more details.

Table 2 - G-506 production distribution by contract and plant (cont.)

9	6	3	2	Chevrolet Plant Code Number
Norwood	Oakland	St. Louis	Tarrytown	Location
				DA-W-398-QM-8
			x	DA-W-398-QM-243
		x		DA-W-398-QM-421
				W-374-ORD-6582
x				DA-W-398-QM-517
				W-374-ORD-1131
				W-374-ORD-1185
				W-374-ORD-1281
x		x		W-374-ORD-2216
x	x	x	x	W-374-ORD-2694
x	x	x	x	W-374-ORD-2694S
				W-374-ORD-2744
				later ORD-2744 W-398-QM-11508
x		x		W-374-ORD-2771
x				W-374-ORD-2797
			x	W-374-ORD-2816
		x		W-374-ORD-2831
x				W-374-ORD-2839
				W-374-ORD-2973
				W-374-ORD-4052
				M6 parts W-374-ORD-5146
				W-374-ORD-5151
				W-374-ORD-6582
				W-398-QM-8278
				W-398-QM-8686
				W-398-QM-8778
				W-398-QM-9402
				W-398-QM-10328
				W-398-QM-10708
				W-398-QM-10939
		x		W-398-QM-11546
				Nos-81721
x				Nos-95703
	x		x	Nos-95878
				Nos-2919

Table 3 - Monthly deliveries as recorded in *Recapitulation of Facility Expansions for Tracked & Wheeled*

Monthly Deliveries 1940

Continues →

	M6 Bomb Service	Cargo w/o winch	Cargo w/ winch	Long Wheelbase	Chassis w/o winch	Dump w/o winch	Dump w/ winch
January	144						
February							
March							
April							
May							
June							
July							
August							
September							
October							
November		1,504	2			6	
December		1,741	136			623	41
Annual Total	144	3,245	138	0	0	629	41
Running Total	144	3,245	138	0	0	629	41

Monthly Deliveries 1941

Continues →

	M6 Bomb Service	Cargo w/o winch	Cargo w/ winch	Long Wheelbase	Chassis w/o winch	Dump w/o winch	Dump w/ winch
January		1,197	1,541	81		256	185
February		1,358	367	109		282	45
March		603	64	28		391	13
April			8		30		
May			1,000				
June			1414			408	11
July			1,434	33	2	738	191
August			1,211	2		203	274
September		172	1,606		31	35	24
October	68	1,924	88		51	66	
November	185	1,634	2,358		52	100	28
December	238	999	1,063	70	23	476	357
Annual Total	491	7,887	12,154	323	189	2,955	1,128
Running Total	635	11,132	12,292	323	189	3,584	1,169

Vehicles from 1940 to July 15, 1945.

Monthly Deliveries 1940 (cont.)

K-44	K-51	K-33 COE	K-42/K-43	Tractor	Chassis for Dump	Monthly Total	
						144	January
						0	February
						0	March
						0	April
						0	May
						0	June
						0	July
						0	August
						0	September
						0	October
						1512	November
	22					2563	December
0	22	0	0	0	0	4219	Annual Total
0	22	0	0	0	0	4219	Running Total

Monthly Deliveries 1941 (cont.)

K-44	K-51	K-33 COE	K-42/K-43	Tractor	Chassis for Dump	Monthly Total	
	12	6				3,278	January
	15	41				2,217	February
	6	36	18			1,159	March
10		13	115			176	April
19			37			1,056	May
8	111		12	53		2,017	June
12	35			25		2,470	July
14				11		1,715	August
18						1,886	September
	10					2,207	October
	78			38		4,473	November
	72			10		3,308	December
81	339	96	182	137	0	25,962	Annual Total
81	361	96	182	137	0	30,181	Running Total

Table 3 - Monthly deliveries as recorded in *Recapitulation of Facility Expansions for Tracked & Wheeled*

Monthly Deliveries 1942

Continues →

	M6 Bomb Service	Cargo w/o winch	Cargo w/ winch	Long Wheelbase	Chassis w/o winch	Dump w/o winch	Dump w/ winch
January	148	845	449	53	5	369	170
February	95	2,771	60	7		117	37
March	128	2,096	276			61	202
April	239	3,065	55		12		215
May	63	2,381	642		101	612	209
June	5	2,215	1,105		249	690	726
July		1,717	1,037		60	388	269
August		1,658	579		111	249	392
September		1,683	734		290	502	850
October		1,238	679		290	294	348
November	206	1,474	442		220	211	234
December	325	2,619	267		236	381	277
Annual Total	1,209	23,762	6,325	60	1,574	3,874	3,929
Running Total	1,844	34,894	18,617	383	1,763	7,458	5,098

Monthly Deliveries 1943

Continues →

	M6 Bomb Service	Cargo w/o winch	Cargo w/ winch	Long Wheelbase	Chassis w/o winch	Dump w/o winch	Dump w/ winch
January	431	1,020	4		40	373	35
February	356	1,312			3	609	
March	483	1,070	1		158	542	
April	276	1,560			13	28	
May	732	1,814			200	8	
June	499	1,160					
July	280	1,732					
August	809	1,830					
September	278	2,624					
October	257	2,172	448				
November	48	1,634	618		281		
December	604	935	884		198		
Annual Total	5,053	18,863	1955	0	893	1,560	35
Running Total	6,897	53,757	20572	383	2,656	9,018	5,133

Vehicles from 1940 to July 15, 1945. **(cont.)**

Monthly Deliveries 1942 (cont.)

K-44	K-51	K-33 COE	K-42/K-43	Tractor	Chassis for Dump	Monthly Total	
	254			24		2,317	January
	27	35				3,149	February
	217	15				2,995	March
	89	10				3,685	April
	461			5		4,474	May
2	490		5	36		5,523	June
30	340		26			3,867	July
25	141		123	10		3,288	August
	323		53	242		4,677	September
34	213	71	9	275		3,451	October
63		136	26	311		3,323	November
	266		2	476		4,849	December
154	2,821	267	244	1379	0	45,598	Annual Total
235	3,182	363	426	1516	0	75,779	Running Total

Monthly Deliveries 1943 (cont.)

K-44	K-51	K-33 COE	K-42/K-43	Tractor	Chassis for Dump	Monthly Total	
7	79	122		69		2,180	January
50	145	26	8	246		2,755	February
49	31	37	17	282		2,670	March
41		25	84	63		2,090	April
56		25	151			2,986	May
20			170			1,849	June
25			175			2,212	July
15			247			2,901	August
16			291	113		3,322	September
15			225	198		3,315	October
15	102		300	135		3,133	November
3	119		299	120		3,162	December
312	476	235	1967	1226	0	32,575	Annual Total
547	3,658	598	393	2,742	0	108,354	Running Total

Table 3 - Monthly deliveries as recorded in *Recapitulation of Facility Expansions for Tracked & Wheeled*

Monthly Deliveries 1944

Continues →

	M6 Bomb Service	Cargo w/o winch	Cargo w/ winch	Long Wheelbase	Chassis w/o winch	Dump w/o winch	Dump w/ winch
January	607	1,578	1356				
February	364	1,618	600		500		
March		2,680	190				
April		2,114	190		500		
May		2,206	190		362		
June		2,280	190		200		
July		2,350	190				
August		2,234	64				
September		2,798	190				
October		2,150	190				
November		2,200	390				
December		2,044	358				
Annual Total	971	26,252	4,098	0	1,562	0	0
Running Total	7,868	80,009	24,670	383	4,218	9,018	5,133

Monthly Deliveries 1945

Continues →

	M6 Bomb Service	Cargo w/o winch	Cargo w/ winch	Long Wheelbase	Chassis w/o winch	Dump w/o winch	Dump w/ winch
January		1,206	314		35		
February		1,084	314				
March		1,158	276				
April		1,208	276				
May		1,302	160				
June		904	57		97		
July							
August							
September							
October							
November							
December							
Annual Total	0	6,862	1,397	0	132	0	0
Running Total	7,868	8,6871	26,067	383	4,350	9,018	5,133

Vehicles from 1940 to July 15, 1945. (cont.)

Monthly Deliveries 1944 (cont.)

K-44	K-51	K-33 COE	K-42/K-43	Tractor	Chassis for Dump	Monthly Total	
43			169	87		3,840	January
41			82	50		3,255	February
			103	56		3,029	March
45			96			2,945	April
74			60			2,892	May
47			82			2,799	June
41			115			2,696	July
41			130			2,469	August
41			131			3,160	September
78			133			2,551	October
73			127			2,790	November
50			109			2,561	December
574	0	0	1,337	193	0	34,987	Annual Total
1,121	3,658	598	3,730	2,935	0	143,341	Running Total

Monthly Deliveries 1945 (cont.)

K-44	K-51	K-33 COE	K-42/K-43	Tractor	Chassis for Dump	Monthly Total	
94			85			1,734	January
47			93			1,538	February
80			91			1,605	March
38			91			1,613	April
105			85			1,652	May
70			54		54	1,236	June
						0	July
						0	August
						0	September
						0	October
						0	November
						0	December
434	0	0	499	0	54	9,378	Annual Total
1,555	3,658	598	4,229	2,935	54	152,719	Running Total

Table 4 - G-506 Production Records by Contract

Table 4 was assembled over a period of years utilizing various US & British Government and Chevrolet documents. None of this information was constructed from documenting surviving vehicles, nor speculation based on surviving vehicles or second hand information.

Regrettably, in some instances the period documents conflict with each other. In most such instances, the information below corresponds with what is recorded in MOST of the surviving documents. When there is only a single source for an entry, that is noted in the footnotes, and for the author at least, single source data is always suspect.

Also – some information is simply wrong. The significant untruth presented below are the many entries that state that various trucks were produced at a Chevrolet plant in Detroit. This information is taken from the *Alphabetic Listing of Major War Supply Contracts, cumulative, June 1940 through September 1945* which was prepared by the Civilian Production Administration, Industrial Statistics Division – hereafter referred to as WSC.

In fact, no such plant existed, and but for experimental vehicles assembled in the company lab, none of these trucks were produced in Detroit. Through 1942 the Experimental Lab assembled 12 vehicles.

However, Detroit was the home of Chevrolet Central Office, and as noted in Chapter 1, the early G-506 trucks were built on Central Office Production Orders. The introduction to the WSC states "An effort has been made to determine the point of final manufacture or assembly and to allocate the individual contracts on this basis rather than on the basis of the headquarters of the contracting company." It goes on to state, "Cases in which the manufacturer or work location has not been identified have been footnoted, with the contract assigned to the company's address." The Detroit contracts lack such a footnote, but the author believes that this was an oversight by those assembling the WSC, or they simply did not know that there was no Detroit manufacturing facility, and thus used the Central Office location as a production location.

Among the many documents used in creating this table, special mention should be made of the *Summary Report of Acceptances Tank-Automotive Material, 1940-1945*, dated 21 January 1946; the *Ordnance Department*

QM Contract #	Ordnance Contract #	Prod Order Number	Bid	IFB Date	Proposal Date	Lettter Order Date	Contract Date	Award Date on WSC
W-398- QM-8278			2-1765	07.11.40	07.19.40		08.06.40	
8278			2-1765					
8278			2-1765					
8278			2-1765					
8278			2-1765					
8278			2-1765					
8278			2-1765					
8278			2-1765					
8278			2-1765					

ITEMS with *: This data point is recorded only in the noted document(s).
ITEMS with **: This data point is recorded only in *Ordnance Department Administrative and Tactical Vehicles 1940-1945, Defense Aid.*

Administrative and Tactical Vehicles 1940-1945; Ordnance Department Administrative and Tactical Vehicles 1940-1945, Defense Aid; and *Recapitulation of Facility Expansions for Tracked & Wheeled Vehicles from 1940 to July 15, 1945*. All four of these documents were prepared by the Office of Chief of Ordnance – Detroit, and are abbreviated in the tables below as SRATAMA, ATV, ATV-DA, and RECAP.

The numerous other data points came from careful examination of Chevrolet internal documents, Technical Manuals and voluminous correspondence in Army files.

How to Read This Table

This table was originally created to aid in the writing of this book. You will notice a great many blank spaces in the table. In most instances, those spaces can be assumed to contain dittos, which were omitted to prevent clutter and aid readability.

For example, in Contract W-398-QM-8278, (the first contract listed), all the trucks, cargo, cargo w/w, dump, etc, had the same Invitation for Bid, Proposal and Contract dates, as listed in the first row of information concerning this contract.

The fourth column from the right lists Acceptance by quantity and year. This information is from the Summary Report of Acceptances Tank-Automotive Material, *1940-1945, and again using the first entry as an example, 55 trucks were ordered, 22 of which were delivered in 1940, and 33 in 1941. When it has been possible to associate US Army registration numbers – the so-called "USA numbers" – to a group of vehicles on a specific contract, that information is shown in the two columns at far right, which are the beginning and ending numbers of the registration number range.*

As stated in the text, production of these trucks began under the auspices of the Quartermaster Corps, later becoming Ordnance responsibility. As such, the contract numbers transitioned. Accordingly, we have parallel columns listing both numbers where applicable. In a few cases, falsely listed by me in the Quartermaster Contract column can be seen contract numbers beginning NOs, which are actually Navy contracts, and a few containing "AC", which are actually Air Corps contracts. There are so few of these, and they are readily distinguishable, I did not feel it warranted the additional space that would be required by adding individual columns for each.

Completion Date on WSC	Year	Model	Plants Involved	Qty Ordered	Accepted per SRTAMA	Acceptance by year qty/yr	Vehicle Type	Start Reg No.	End Reg No.
	1941	G-4112YP	Detroit	54	55	22-1940	Panel delivery by SRTAMA*	328401	328442
			Detroit			33-1941	Panel delivery by SRTAMA*	331684	331695
	1941	G-4112YP	Detroit	6403	6403	3245-1940	cargo	320560	325073
			Detroit			3158-1941	cargo	328951	330839
	1941	G-4112YP	Detroit	2110	2110	138-1940	cargo w/w	325074	326802
			Detroit			1972-1941			
			Detroit				cargo w/w	330840	330965
			Detroit				cargo w/w	326803	327057
	1941	G-4112YQ	Detroit	219	218	218-1941	LWB cargo	328310	328400
			Detroit				LWB cargo	331556	331683

ITEMS with *: This data point is recorded only in the noted document(s).
ITEMS with **: This data point is recorded only in *Ordnance Department Administrative and Tactical Vehicles 1940-1945, Defense Aid.*

Table 4 - G-506 Production Records by Contract (cont.)

Continues →

QM Contract #	Ordnance Contract #	Prod Order Number	Bid	IFB Date	Proposal Date	Lettter Order Date	Contract Date	Award Date on WSC
8278			2-1765					
8278			2-1765					
8278			2-1765					
W-398-QM-8778			2-1996	09.09.40	09.26.40		11.05.40	
8778			2-1996					
8778			2-1996					
8778			2-1996					
8778			2-1996					
W-398-QM-8686			2-2021	09.16.40	09.24.40		10.24.40	
8686								
8686	W-535-AC-16747			10.11.40	10.17.40		12.10.40	11.01.40
W-398-QM-9250			2-2892	10.30.40	01.09.41		01.13.41	
NOs-81721				11.25.40	12.16.40		02.10.41	
W-398-QM-9402			2-2810	12.20.40	12.21.40		01.23.41	
9402			2-2810					
9402			2-2810					
9402			2-2810					
9402			2-2810					
9402								
9402			2-2810					
9402			2-2810					
9402			2-2810					

ITEMS with *: This data point is recorded only in the noted document(s).
ITEMS with **: This data point is recorded only in *Ordnance Department Administrative and Tactical Vehicles 1940-1945, Defense Aid.*

Completion Date on WSC	Year	Model	Plants Involved	Qty Ordered	Accepted per SRTAMA	Acceptance by year qty/yr	Vehicle Type	Start Reg No.	End Reg No.
	1941	G-4112YP	Detroit	1558	1558	629-1940	dump	327058	328025
			Detroit			929-1941	dump	330966	331555
			Detroit	284	284	41-1940	dump w/w	328026	328309
			Detroit			243-1941			
	1941	G-4112YP	Detroit	182	182	182-1941	tele. Maint. (67 wo/w)	002968	003034
			Detroit					002852	single
			Detroit					002854	002967
	1941	G-4112YP	Detroit	81			K-44	002853	single
			Detroit					003035	003114
	1941	G-4103YX	Detroit	96	96	96-1941	COE stake and platform K-33	328665	328760
	1941	G-4112YP	Detroit	8			tractor truck	328761	328768
03.01.41			American Gas Accumulator Co Elizabeth NJ				field lighting		
	1941	G-4112YP	ATV & SRTAMA*	30	30	30-1941	airfield service chassis by STRAMA*	50702	50731
			Detroit				NAVY		
	1941	G-4163ZP	Detroit	200		6769-1941	cargo w/w for Defense Aid**		
			Detroit			1-1942			
	1941	G-4163ZP	Detroit	6804	6770	1-1942	cargo w/w	332065	single
			Detroit			6769-1942			
			Detroit				cargo w/w	332066	335851
			Detroit				cargo w/w	339995	342977
			Detroit				cargo w/w	335852	335886
			Detroit		35	35-1941	LWB cargo W/W		
	1941	G-4105ZP	Detroit	146	146	146-1941	panel delivery	337301	337446
	1941	G-4152ZP	Detroit	1384	1414	1384-1941	dump	335887	337270
	1941	G-7106NH	Detroit	30			dump	337271	337300

ITEMS with *: This data point is recorded only in the noted document(s).
ITEMS with **: This data point is recorded only in *Ordnance Department Administrative and Tactical Vehicles 1940-1945, Defense Aid.*

Table 4 - G-506 Production Records by Contract (cont.)

Continues →

QM Contract #	Ordnance Contract #	Prod Order Number	Bid	IFB Date	Proposal Date	Lettter Order Date	Contract Date	Award Date on WSC
9402			2-2810					
9402			2-2810					
9402			2-2810					
9402			2-2810					
9402			2-2810					
DA-W-398-QM-8			2-3990	05.06.41	05.14.41		06.07.41	
8								
8								
8								
8								
8								
W-398-QM-10328	W-374-ORD-2726		2-3990	05.06.41	05.14.41		06.07.41	
10328			2-3990					
10328			2-3990					
10328			2-3990					
10328			2-3990					
10328			2-3990					
10328			2-3990					

ITEMS with *: This data point is recorded only in the noted document(s).
ITEMS with **: This data point is recorded only in *Ordnance Department Administrative and Tactical Vehicles 1940-1945, Defense Aid.*

Completion Date on WSC	Year	Model	Plants Involved	Qty Ordered	Accepted per SRTAMA	Acceptance by year qty/yr	Vehicle Type	Start Reg No.	End Reg No.
	1941	G-4162ZP	Detroit	500	500	500-1941	dump w/w	332064	single
			Detroit				dump w/w	337447	337945
	1941	G-4165ZP	Detroit		89	89-1941	tractor	337946	338034
	1941	G-4113ZP	Detroit	2	3	3-1941	cab and chassis	607641	607642
	1941	G-4113ZP	Detroit	1			Oil servicing	80920	single
	1942	G-7107NJ	Detroit	24	24	24-1941	cargo for Defense Aid**		
	1942	G-7113NK	Detroit	24	24	24-1942	tractor for Defense Aid**		
	1942	G-7117NM	Detroit	2500	2500	2500-1941	cargo w/w for Defense Aid**		
	1942	G-7127NP	Detroit	50	50	50-1941	LWB cargo W/W for Defense Aid**		
	1942	G-7127NL	Detroit	350	350	256-1941	dump w/w for Defense Aid**		
			Detroit			94-1942			
			Detroit		180	180-1942	panel delivery**		
	1942	G-7105NG	Detroit	86	86	13-1941	panel delivery	347683	347707
			Detroit			73-1942	panel delivery	348821	348840
			Detroit				panel delivery	353189	353228
			Detroit				panel delivery	359216	single
	1942	G-7107NJ	Detroit	7086	7627		cargo	345641	347272
			Detroit			4417-1941			
			Detroit			2372-1942			
			Detroit			838-1943			
			Detroit				cargo	347741	348740
			Detroit				cargo	349739	350393

ITEMS with *: This data point is recorded only in the noted document(s).
ITEMS with **: This data point is recorded only in *Ordnance Department Administrative and Tactical Vehicles 1940-1945, Defense Aid.*

Table 4 - G-506 Production Records by Contract (cont.)

Continues →

QM Contract #	Ordnance Contract #	Prod Order Number	Bid	IFB Date	Proposal Date	Lettter Order Date	Contract Date	Award Date on WSC
10328			2-3990					
10328			2-3990					
10328			2-3990					
10328			2-3990					
10328			2-3990					
10328			2-3990					
10328			2-3990					
10328			2-3990					
10328			2-3990					
10328			2-3990					
10328			2-3990					
10328			2-3990					
10328			2-3990					
10328			2-3990					
10328			2-3990					
10328			2-3990					
10328			2-3990					
10328			2-3990					
10328			2-3990					
10328			2-3990					
10328			2-3990					
10328			2-3990					
10328			2-3990					
10328			2-3990					
10328			2-3990					
10328			2-3990					
10328			2-3990					
10328			2-3990					
10328			2-3990					
10328			2-3990					

ITEMS with *: This data point is recorded only in the noted document(s).
ITEMS with **: This data point is recorded only in *Ordnance Department Administrative and Tactical Vehicles 1940-1945, Defense Aid.*

Completion Date on WSC	Year	Model	Plants Involved	Qty Ordered	Accepted per SRTAMA	Acceptance by year qty/yr	Vehicle Type	Start Reg No.	End Reg No.
			Detroit				cargo	350769	351488
			Detroit				cargo	348761	348810
			Detroit				cargo	348841	349630
			Detroit				cargo	351769	352777
			Detroit				cargo	356916	357465
			Detroit				cargo	357651	358300
			Detroit				cargo	358641	358940
	1942	G-7127NP	Detroit	80	80	20-1941	LWB	347423	347442
			Detroit			60-1942			
			Detroit				LWB	348741	348760
			Detroit				LWB	352934	352973
	1942	G-7117NM	Detroit	1200		441-1941	cargo w/w	347273	347422
			Detroit			218-1942			
			Detroit				cargo w/w	349411	349459
			Detroit				cargo w/w	350394	350513
			Detroit				cargo w/w	351489	351618
			Detroit				cargo w/w	352784	352933
			Detroit				cargo w/w	353244	single
			Detroit				cargo w/w	357466	357615
			Detroit				cargo w/w	358301	358550
			Detroit				cargo w/w	358941	359140
	1942	G-7106NH	Detroit	1150	1150	642-1941	dump	347443	347642
			Detroit			508-1942			
			Detroit				dump	348761	348810
			Detroit				dump	349460	349709
			Detroit				dump	350514	350713
			Detroit				dump	351619	351718
			Detroit				dump	352974	353173
			Detroit				dump	357616	357650
			Detroit				dump	358551	358615
			Detroit				dump	359141	359190
	1942	G-7116NL	Detroit	225	250	129-1941	dump w/w	347643	347682
			Detroit			121-1942			
			Detroit				dump w/w	348811	348820

ITEMS with *: This data point is recorded only in the noted document(s).
ITEMS with **: This data point is recorded only in *Ordnance Department Administrative and Tactical Vehicles 1940-1945, Defense Aid.*

Table 4 - G-506 Production Records by Contract (cont.)

Continues →

QM Contract #	Ordnance Contract #	Prod Order Number	Bid	IFB Date	Proposal Date	Lettter Order Date	Contract Date	Award Date on WSC
10328			2-3990					
10328			2-3990					
10328			2-3990					
10328			2-3990					
10328			2-3990					
10328			2-3990					
10328			2-3990					
10328			2-3990					
10328			2-3990					
10328			2-3990					
	W-374-ORD-1131	T-1510	2-3945	05.20.41	05.24.41	06.07.41	09.29.41	10.01.42
W-398-QM-10708			2-4502 or 4562	06.17.41	06.23.41		06.29.41	
W-535-AC-19340 (AGA)			2-2773	08.12.41	09.04.41		10.02.41	06.01.41
W-398-QM-10939	W-374-ORD-2733		2-4921	08.12.41	09.02.41		10.13.41	
10939			2-4921					
10939			2-4921					
10939								
10939								
	W-374-ORD-1185	T-2068	2-3945	09.02.41	09.05.41		09.13.41	01.01.42
	1185	T-2276						02.01.42

ITEMS with *: This data point is recorded only in the noted document(s).
ITEMS with **: This data point is recorded only in *Ordnance Department Administrative and Tactical Vehicles 1940-1945, Defense Aid.*

Completion Date on WSC	Year	Model	Plants Involved	Qty Ordered	Accepted per SRTAMA	Acceptance by year qty/yr	Vehicle Type	Start Reg No.	End Reg No.
			Detroit				dump w/w	349710	349738
			Detroit				dump w/w	350714	350768
			Detroit				dump w/w	351719	351768
			Detroit				dump w/w	353174	353188
			Detroit				dump w/w	353243	single
			Detroit				dump w/w	358616	358640
	1942	G-7113NK	Detroit	48	48	48-1941	tractor	347708	347740
			Detroit				tractor	352783	single
			Detroit				tractor	353229	353242
			Detroit		161	156-1941	cab and chassis		
			Detroit			5-1942			
	1942	G-7133NZ	Detroit	126			airfield crash truck	501304	501429
12.01.42	1941	G-7128NQ	Flint	491	491	491-1941	bomb service	004373	004863
	1942	G-7123NN	Baltimore	60	60	60-1942	COE stake and platform K-33	343799	343838
12.01.41		G-7143NF	American Gas Accumulator Co Elizabeth NJ	142			field lighting	609058	609199
	1942	G-7107NJ	Flint	1627	2482	1300-1942	cargo	359275	360901
			Flint			1182-1943			
	1942	G-7117NM	Flint	1225	213	213-1942	cargo w/w	360902	362126
	1942	G-7116NL	Flint	500	500	500-1942	dump w/w	362127	362626
			Flint		113	113-1942	cab and chassis wo/w		
			Flint		44	44-1942	tractor		
01.01.42	1943	G-7128NQ	Baltimore	116		243-1942	bomb service	006070	006185
02.01.42	1943	G-7128NQ	Baltimore	127	243	243-1942	bomb service	006186	006312

ITEMS with *: This data point is recorded only in the noted document(s).
ITEMS with **: This data point is recorded only in *Ordnance Department Administrative and Tactical Vehicles 1940-1945, Defense Aid.*

Table 4 - G-506 Production Records by Contract (cont.)

Continues →

QM Contract #	Ordnance Contract #	Prod Order Number	Bid	IFB Date	Proposal Date	Lettter Order Date	Contract Date	Award Date on WSC
W-398-QM-10912	W-374-ORD-2993		2-4996	09.15.41	09.18.41		09.20.41	
WA-W-398-QM-171	W-374-ORD-2964		2-5024	09.26.41	10.06.41		11.13.41	
Nos-95703			2-5160	10.27.41	11.12.41		01.10.42	01.01.42
W-535-AC-888 L (AGA)			2-5243	11.10.41	12.08.41		12.11.41	12.01.41
W-398-QM-11508	W-374-ORD-2744	T-5685	2-5272	11.18.41	12.04.41		01.05.42	
11508		T-5685	2-5272					
11508		T-5685	2-5272					
11508		T-5685	2-5272					
W-398-QM-11546	W-374-ORD-2748	T-5881	2-5469	12.12.41	12.13.41		02.06.42	
	W-374-ORD-1281	T-3114	2-5487	12.19.41	12.22.41		01.15.42	03.01.42
W-398-QM-13383	W-374-ORD-2795	T-5687	2-5615	02.07.42	02.24.42	05.18.42	08.31.42	
	W-374-ORD-1580	T-3657		02.13.42	02.27.42		07.15.42	
	1580	T-3952						

ITEMS with *: This data point is recorded only in the noted document(s).
ITEMS with **: This data point is recorded only in *Ordnance Department Administrative and Tactical Vehicles 1940-1945, Defense Aid.*

Completion Date on WSC	Year	Model	Plants Involved	Qty Ordered	Accepted per SRTAMA	Acceptance by year qty/yr	Vehicle Type	Start Reg No.	End Reg No.
	1942	G-7105NG	Chevy doc & SRTAMA & ATV*	180	180	147-1941	panel delivery	356736	356915
			Chevy doc & SRTAMA & ATV*			33-1942			
	1942	G-7116NL	Chevy doc & SRTAMA &ATV-DA*	16	16	16-1942	tractor** (DUMP w/w by SRTAMA)* w/w for Defense Aid**		
10.01.42		G-7153NX	Baltimore, Norwood				COE Fire Truck for NAVY		
12.01.42	1942	G-7143NF	American Gas Accumulator Co Elizabeth NJ	100			field lighting for Defense Aid**		
	1942	G-7173NS	Cleveland	181	207	207-1942	K-43 w/w	007213	007393
	1942	G-7173NS	Cleveland	26			K-43 w/w	007932	007957
	1942	G-7163NR	Cleveland	154	154	154-1942	K-44	007394	007518
			Cleveland				K-44	007958	007986
	1942	G-7105NG	St. Louis	217	217	217-1942	panel delivery	365531	365747
03.01.42	1943	G-7128NQ	Baltimore	435	435	435-1942	bomb service	007987	008421
			Norwood	357	357	207-1942	COE stake and platform K-33	3114446	3114802
			Norwood			150-1943	COE stake and platform K-34		
	1943	G-7128NQ	SRTAMA, ATV*	546	548	531-1942	bomb service	0012010	0012555
			SRTAMA, ATV*			17-1943	bomb service		
	1943	G-7128NQ	SRTAMA, ATV*	1		?	bomb service	0012557	single

ITEMS with *: This data point is recorded only in the noted document(s).
ITEMS with **: This data point is recorded only in *Ordnance Department Administrative and Tactical Vehicles 1940-1945, Defense Aid.*

Table 4 - G-506 Production Records by Contract (cont.)

Continues →

QM Contract #	Ordnance Contract #	Prod Order Number	Bid	IFB Date	Proposal Date	Lettter Order Date	Contract Date	Award Date on WSC
W-398-QM-11510	W-374-ORD-2694		2-5528	03.05.42	03.30.42	12.31.41	01.04.42	01.01.42
11510		T-6057	2-5528					
11510		T-6057	2-5528					
11510		T-6057	2-5528					
11510		T-6057	2-5528					
11510		T-6057	2-5528					

ITEMS with *: This data point is recorded only in the noted document(s).
ITEMS with **: This data point is recorded only in *Ordnance Department Administrative and Tactical Vehicles 1940-1945, Defense Aid.*

Completion Date on WSC	Year	Model	Plants Involved	Qty Ordered	Accepted per SRTAMA	Acceptance by year qty/yr	Vehicle Type	Start Reg No.	End Reg No.
12/1/1943*			Atlanta, Baltimore, Norwood, Oakland, St. Louis, Tarrytown						
	1942	G-7123NN	Atlanta, Baltimore, Norwood, Oakland, St. Louis, Tarrytown	85			COE stake and platform K-33	3219414	3219498
	1942	G-7107NJ	Atlanta, Baltimore, Norwood, Oakland, St. Louis, Tarrytown	30705	30705	14624-1942	cargo	371746	388973
			Atlanta, Baltimore, Norwood, Oakland, St. Louis, Tarrytown			15893-1943			
			Atlanta, Baltimore, Norwood, Oakland, St. Louis, Tarrytown			188-1944			
			Atlanta, Baltimore, Norwood, Oakland, St. Louis, Tarrytown				cargo	388974	389178
			Atlanta, Baltimore, Norwood, Oakland, St. Louis, Tarrytown				cargo	390024	390268
			Atlanta, Baltimore, Norwood, Oakland, St. Louis, Tarrytown				cargo	393514	393868

ITEMS with *: This data point is recorded only in the noted document(s).
ITEMS with **: This data point is recorded only in *Ordnance Department Administrative and Tactical Vehicles 1940-1945, Defense Aid.*

Table 4 - G-506 Production Records by Contract (cont.)

Continues →

QM Contract #	Ordnance Contract #	Prod Order Number	Bid	IFB Date	Proposal Date	Lettter Order Date	Contract Date	Award Date on WSC
11510		T-6057	2-5528					
11510		T-6057	2-5528					
11510		T-6057	2-5528					
11510		T-6057	2-5528					
11510		T-6057	2-5528					
11510		T-6057	2-5528					
11510		T-6057	2-5528					

ITEMS with *: This data point is recorded only in the noted document(s).
ITEMS with **: This data point is recorded only in *Ordnance Department Administrative and Tactical Vehicles 1940-1945, Defense Aid.*

Completion Date on WSC	Year	Model	Plants Involved	Qty Ordered	Accepted per SRTAMA	Acceptance by year qty/yr	Vehicle Type	Start Reg No.	End Reg No.
			Atlanta, Baltimore, Norwood, Oakland, St. Louis, Tarrytown				cargo	394869	395621
			Atlanta, Baltimore, Norwood, Oakland, St. Louis, Tarrytown				cargo	396002	398927
			Atlanta, Baltimore, Norwood, Oakland, St. Louis, Tarrytown				cargo	3101874	3101883
			Atlanta, Baltimore, Norwood, Oakland, St. Louis, Tarrytown				cargo	3218319	3218878
			Atlanta, Baltimore, Norwood, Oakland, St. Louis, Tarrytown				cargo	3259288	3259337
			Atlanta, Baltimore, Norwood, Oakland, St. Louis, Tarrytown				cargo	3284244	3292616
	1942	G-7117NM	Atlanta, Baltimore, Norwood, Oakland, St. Louis, Tarrytown	8650	8650	5465-1942	cargo w/w	389179	390025
			Atlanta, Baltimore, Norwood, Oakland, St. Louis, Tarrytown			1955-1943			

ITEMS with *: This data point is recorded only in the noted document(s).
ITEMS with **: This data point is recorded only in *Ordnance Department Administrative and Tactical Vehicles 1940-1945, Defense Aid*.

Table 4 - G-506 Production Records by Contract (cont.)

Continues →

QM Contract #	Ordnance Contract #	Prod Order Number	Bid	IFB Date	Proposal Date	Lettter Order Date	Contract Date	Award Date on WSC
11510		T-6057	2-5528					
11510		T-6057	2-5528					
11510		T-6057	2-5528					
11510		T-6057	2-5528					
11510		T-6057	2-5528					
11510		T-6057	2-5528					

ITEMS with *: This data point is recorded only in the noted document(s).
ITEMS with **: This data point is recorded only in *Ordnance Department Administrative and Tactical Vehicles 1940-1945, Defense Aid.*

Completion Date on WSC	Year	Model	Plants Involved	Qty Ordered	Accepted per SRTAMA	Acceptance by year qty/yr	Vehicle Type	Start Reg No.	End Reg No.
			Atlanta, Baltimore, Norwood, Oakland, St. Louis, Tarrytown			1230-1943			
			Atlanta, Baltimore, Norwood, Oakland, St. Louis, Tarrytown				cargo w/w	390269	393513
			Atlanta, Baltimore, Norwood, Oakland, St. Louis, Tarrytown				cargo w/w	393869	394868
			Atlanta, Baltimore, Norwood, Oakland, St. Louis, Tarrytown				cargo w/w	395622	396001
			Atlanta, Baltimore, Norwood, Oakland, St. Louis, Tarrytown				cargo w/w	3293096	3296275
	1942	G-7105NG	Atlanta, Baltimore, Norwood, Oakland, St. Louis, Tarrytown	1169	1169	693-1942	panel delivery	3106797	3107021
			Atlanta, Baltimore, Norwood, Oakland, St. Louis, Tarrytown			476-1943			
			Atlanta, Baltimore, Norwood, Oakland, St. Louis, Tarrytown				panel delivery	3103417	3103813

ITEMS with *: This data point is recorded only in the noted document(s).
ITEMS with **: This data point is recorded only in *Ordnance Department Administrative and Tactical Vehicles 1940-1945, Defense Aid.*

Table 4 - G-506 Production Records by Contract (cont.)

Continues →

QM Contract #	Ordnance Contract #	Prod Order Number	Bid	IFB Date	Proposal Date	Lettter Order Date	Contract Date	Award Date on WSC
11510		T-6057	2-5528					
11510		T-6057	2-5528					
11510			2-5528					
11510			2-5528					
11510			2-5528					
11510		T-6057	2-5528					
11510		T-6057	2-5528					

ITEMS with *: This data point is recorded only in the noted document(s).
ITEMS with **: This data point is recorded only in *Ordnance Department Administrative and Tactical Vehicles 1940-1945, Defense Aid.*

Completion Date on WSC	Year	Model	Plants Involved	Qty Ordered	Accepted per SRTAMA	Acceptance by year qty/yr	Vehicle Type	Start Reg No.	End Reg No.
			Atlanta, Baltimore, Norwood, Oakland, St. Louis, Tarrytown				panel delivery	3105514	3105654
			Atlanta, Baltimore, Norwood, Oakland, St. Louis, Tarrytown				panel delivery	3218879	3219284
	1942	G-7107NJ	Atlanta, Baltimore, Norwood, Oakland, St. Louis, Tarrytown	6300			cargo for Defense Aid**		
	1942	G-7117NM	Atlanta, Baltimore, Norwood, Oakland, St. Louis, Tarrytown	2			cargo w/w for Defenese Aid**		
	1942	G-7116NL	Atlanta, Baltimore, Norwood, Oakland, St. Louis, Tarrytown	4			dump w/w for Defense Aid**		
	1942	G-7106NH	Atlanta, Baltimore, Norwood, Oakland, St. Louis, Tarrytown	4886	4886	3326-1942	dump wo/w	398928	399999
			Atlanta, Baltimore, Norwood, Oakland, St. Louis, Tarrytown			1560-1943			
			Atlanta, Baltimore, Norwood, Oakland, St. Louis, Tarrytown				dump wo/w	3100000	3101873

ITEMS with *: This data point is recorded only in the noted document(s).
ITEMS with **: This data point is recorded only in *Ordnance Department Administrative and Tactical Vehicles 1940-1945, Defense Aid.*

Table 4 - G-506 Production Records by Contract (cont.)

Continues →

QM Contract #	Ordnance Contract #	Prod Order Number	Bid	IFB Date	Proposal Date	Lettter Order Date	Contract Date	Award Date on WSC
11510		T-6057	2-5528					
11510		T-6057	2-5528					
11510		T-6057	2-5528					
11510		T-6057	2-5528					
11510		T-6057	2-5528					

ITEMS with *: This data point is recorded only in the noted document(s).
ITEMS with **: This data point is recorded only in *Ordnance Department Administrative and Tactical Vehicles 1940-1945, Defense Aid.*

Completion Date on WSC	Year	Model	Plants Involved	Qty Ordered	Accepted per SRTAMA	Acceptance by year qty/yr	Vehicle Type	Start Reg No.	End Reg No.
			Atlanta, Baltimore, Norwood, Oakland, St. Louis, Tarrytown				dump wo/w	3121234	3123173
	1942	G-7116NL	Atlanta, Baltimore, Norwood, Oakland, St. Louis, Tarrytown	1700	3233	3198-1942	dump w/w	3103814	3105513
			Atlanta, Baltimore, Norwood, Oakland, St. Louis, Tarrytown			35-1943			
	1942	G-7116NL	Atlanta, Baltimore, Norwood, Oakland, St. Louis, Tarrytown	1533			dump w/w	3101884	3103416
	1942	G-7173NS	Atlanta, Baltimore, Norwood, Oakland, St. Louis, Tarrytown	2463	2463	37-1942	K-43 w/w	0016555	0016591
			Atlanta, Baltimore, Norwood, Oakland, St. Louis, Tarrytown			1967-1943			
			Atlanta, Baltimore, Norwood, Oakland, St. Louis, Tarrytown			459-1944			
	1942	G-7163NR	Atlanta, Baltimore, Norwood, Oakland, St. Louis, Tarrytown	522	522	312-1943	K-44	0016592	0016794

ITEMS with *: This data point is recorded only in the noted document(s).
ITEMS with **: This data point is recorded only in *Ordnance Department Administrative and Tactical Vehicles 1940-1945, Defense Aid.*

Table 4 - G-506 Production Records by Contract (cont.)

Continues →

QM Contract #	Ordnance Contract #	Prod Order Number	Bid	IFB Date	Proposal Date	Lettter Order Date	Contract Date	Award Date on WSC
11510		T-6057	2-5528					
11510		T-6057	2-5528					
11510		T-6057	2-5528					
11510		T-6057	2-5528					
11510		T-6057	2-5528					
11510		T-6057	2-5528					
11510		T-6057						

ITEMS with *: This data point is recorded only in the noted document(s).
ITEMS with **: This data point is recorded only in *Ordnance Department Administrative and Tactical Vehicles 1940-1945, Defense Aid.*

Completion Date on WSC	Year	Model	Plants Involved	Qty Ordered	Accepted per SRTAMA	Acceptance by year qty/yr	Vehicle Type	Start Reg No.	End Reg No.
			Atlanta, Baltimore, Norwood, Oakland, St. Louis, Tarrytown			210-1944	K-44	0032262	0032580
	1944	G-7163NR	Atlanta, Baltimore, Norwood, Oakland, St. Louis, Tarrytown				K-44	0080856	0081142
	1942	G-7113NK	Atlanta, Baltimore, Norwood, Oakland, St. Louis, Tarrytown	2604	2604	1185-1942	tractor	3105655	3106796
			Atlanta, Baltimore, Norwood, Oakland, St. Louis, Tarrytown			1226-1943	tractor	3107022	3107123
			Atlanta, Baltimore, Norwood, Oakland, St. Louis, Tarrytown			193-1944	tractor	3219499	3220099
			Atlanta, Baltimore, Norwood, Oakland, St. Louis, Tarrytown				tractor	3296276	3297034
			Atlanta, Baltimore, Norwood, Oakland, St. Louis, Tarrytown	1900	1900	1107-1942	cab and chassis wo/w		
			Atlanta, Baltimore, Norwood, Oakland, St. Louis, Tarrytown			793-1943			

ITEMS with *: This data point is recorded only in the noted document(s).
ITEMS with **: This data point is recorded only in *Ordnance Department Administrative and Tactical Vehicles 1940-1945, Defense Aid.*

Table 4 - G-506 Production Records by Contract (cont.)

Continues →

QM Contract #	Ordnance Contract #	Prod Order Number	Bid	IFB Date	Proposal Date	Lettter Order Date	Contract Date	Award Date on WSC
11510			2-5528					
11510			2-5528					
W-398-QM-12980	W-374-ORD-2787	T-5686	2-5638	03.07.42	03.21.42	04.18.42	05.28.42	
W-398-QM-12945	W-374-ORD-2785		2-5637	03.03.42	03.21.42	04.29.42	05.05.42	
W-398-QM-13532	W-374-ORD-2797	T-5375	2-5649	03.21.42	03.30.42		05.18.42	11.01.42
13532		T-5375						
13532								
13532								
W-NOS-95055 (AGA)			2-5243*	04.21.42	05.02.42		05.06.42	12.01.41
	W-374-ORD-2216	T-4294	2-5669	08.01.42	08.21.42	08.01.42	08.01.42	08.01.42
	W-374-ORD-4052	T-5790	2-5676	10.13.42	12.01.42		12.10.42	
	W-374-ORD-5151	T-7136		03.09.43	07.29.43	03.10.43	07.28.43	
	W-374-ORD-5072			05.12.43	10.01.43	05.11.43	01.03.44	
W-398-QM-1096	W-374-ORD-3032							
W-398-QM-12358	W-374-ORD-2722							

ITEMS with *: This data point is recorded only in the noted document(s).
ITEMS with **: This data point is recorded only in *Ordnance Department Administrative and Tactical Vehicles 1940-1945, Defense Aid.*

Completion Date on WSC	Year	Model	Plants Involved	Qty Ordered	Accepted per SRTAMA	Acceptance by year qty/yr	Vehicle Type	Start Reg No.	End Reg No.
	1942	G-7103NE	Atlanta, Baltimore, Norwood, Oakland, St. Louis, Tarrytown	1421			chassis	6061126	6062233
			Atlanta, Baltimore, Norwood, Oakland, St. Louis, Tarrytown				chassis	6092653	6092965
	1945	G-7113NK	ATV & Chevy doc*	126	126	126-1942	tractor	3109758	3109883
	1945	G-7103NE	SRTAMA*	199	199	199-1942	chassis	6044503	6044701
04.01.43	1942	G-7103NE*	Norwood	59	250	150-1942	crash truck chassis		
	1942	G-7103NE*	Norwood	191		100-1943	crash truck chassis	503769	503918
			Norwood				crash truck chassis	503920	single
			Norwood				crash truck chassis	503961	504000
05.01.42		G-7143NF	American Gas Accumulator Co Elizabeth NJ	71			J-3 field lighting 6 for Army, 65 for Navy		
05.01.43	1943	G-7128NQ	Norwood, St. Louis	1701	1701	1701-1943	bomb service	0022800	0024500
	1943	G-7128NQ	Flint	2359	2339	2339-1943	bomb service	0036252	0038610
	1943	G-7128NQ	Detroit		1947	976-1943	bomb service	0061361	0063307
	1943		Detroit			971-1944	bomb service	0061361	0063307
			Chevy doc only*						
	1942	G-7117NM	ATV*	1			cargo w/w	3257837	single
				2			2 experi-mental low silhoutte	3114161	3114162

ITEMS with *: This data point is recorded only in the noted document(s).
ITEMS with **: This data point is recorded only in *Ordnance Department Administrative and Tactical Vehicles 1940-1945, Defense Aid.*

Table 4 - G-506 Production Records by Contract (cont.)

Continues →

QM Contract #	Ordnance Contract #	Prod Order Number	Bid	IFB Date	Proposal Date	Lettter Order Date	Contract Date	Award Date on WSC
W-398-QM-12396	W-374-ORD-2771		2-5584	01.21.42	01.22.42	03.16.42	12.17.42	03.01.42
12396			2-5584					
DA-W-398-QM-243	W-374-ORD-2973		2-5332	11.29.41	12.06.41		01.12.42	12.01.41
DA-W-398-QM-252	W-374-ORD-2816	T-5753	and	12.01.41	12.06.41		01.29.42	11.01.42
252								
252								
252								
252								
DA-W-398-QM-421	W-374-ORD-2831	T-5888	2-5584	01.21.42	01.22.42	03.16.42	03.17.42	03.01.42
	W-20-018-ORD-517			10.05.43	10.18.43		10.25.43	
DA-W-398-QM-517	W-374-ORD-2839	T 5690	2-5627	02.23.42	03.04.42	04.09.42	08.31.42	05.01.42
517		T-14170						
517		T-10511						

ITEMS with *: This data point is recorded only in the noted document(s).
ITEMS with **: This data point is recorded only in *Ordnance Department Administrative and Tactical Vehicles 1940-1945, Defense Aid.*

Completion Date on WSC	Year	Model	Plants Involved	Qty Ordered	Accepted per SRTAMA	Acceptance by year qty/yr	Vehicle Type	Start Reg No.	End Reg No.
09.01.43	1942	G-7105NG	Baltimore, Norwood, St. Louis	1330	1330	1330-1942	panel delivery	370051	370730
								3114813	3115462
07.01.43	1942	G-7105NG	Tarrytown, Flint*	75	75	75-1942	panel for Defense Aid**	3115524	3115598
04.01.43	1942	G-7107NJ	Tarrytown	6458	5754	288-1941	cargo for Defense Aid**		
			Tarrytown			5466-1942			
			Tarrytown		900	472-1941	cargo w/w*		
			Tarrytown			428-1942	cargo w/w*		
	1942	G-7107NJ	Tarrytown	39		?	cargo for Defense Aid**	3109523	3109561
	1942	G-7103NE	Tarrytown	112		?	cargo for Defense Aid**		
	1942	G-7107NJ	Tarrytown	6		?	cargo for Defense Aid**	3344815	3344820
	1942	G-7113NK	Tarrytown	39			tractor for Defense Aid**		
04.01.43	1942	G-7105NG	St. Louis	220	220	220-1942	panel for Defense Aid**		
			Chevy doc only*						
04.01.43	1942	G-7107NJ	Atlanta, Baltimore, Flint, Norwood	950	950	950-1943	cargo for Defense Aid**	3160000	3160949
	1945	G-7107NJ	Atlanta, Baltimore, Flint, Norwood	752			cargo for Defense Aid**	3519012	3519763
	1944	G-7107NJ	Atlanta, Baltimore, Flint, Norwood	15088	15088	15088-1944	cargo	3440079	3455166

ITEMS with *: This data point is recorded only in the noted document(s).
ITEMS with **: This data point is recorded only in *Ordnance Department Administrative and Tactical Vehicles 1940-1945, Defense Aid.*

Table 4 - G-506 Production Records by Contract (cont.)

Continues →

QM Contract #	Ordnance Contract #	Prod Order Number	Bid	IFB Date	Proposal Date	Lettter Order Date	Contract Date	Award Date on WSC
517		T-10543						
517		T-14170						
517		T-10506						
517		T-14168						
517		T-10509						
517		T-15389						
517		T-15389						
517		T-14167						

ITEMS with *: This data point is recorded only in the noted document(s).
ITEMS with **: This data point is recorded only in *Ordnance Department Administrative and Tactical Vehicles 1940-1945, Defense Aid.*

Completion Date on WSC	Year	Model	Plants Involved	Qty Ordered	Accepted per SRTAMA	Acceptance by year qty/yr	Vehicle Type	Start Reg No.	End Reg No.
	1944	G-7107NJ	Atlanta, Baltimore, Flint, Norwood	1065	2427	1526-1944	cargo for Corps of Engineers		
			Atlanta, Baltimore, Flint, Norwood			901-1945	1362 CHASSIS on contract accepted as cargo		
	1944	G-7107NJ	Atlanta, Baltimore, Flint, Norwood	5918	5450	5450-1945	cargo including 468 cab and chassis	3480357	3486274
			Atlanta, Baltimore, Flint, Norwood		468	468-1945	468 cab and chassis		
	1944	G-7117NM	Atlanta, Baltimore, Flint, Norwood	2460	2460	924-1944	cargo w/w	3437343	3439802
			Atlanta, Baltimore, Flint, Norwood			1536-1945			
		G-7107NJ	Atlanta, Baltimore, Flint, Norwood	511	511	511-1945	cargo wo/w (see cab and chassis)	3506501	3507046
	1944	G-7173NS	Atlanta, Baltimore, Flint, Norwood	842	842	842-1944	K-43	0080014	0080855
	1944	G-7173NS	Atlanta, Baltimore, Flint, Norwood	535	535	36-1944	K-43	0097813	0098184
	1945	G-7173NS	Atlanta, Baltimore, Flint, Norwood			499-1945	K-43	0099777	0099939
	1944	G-7163NR	Atlanta, Baltimore, Flint, Norwood	615	615	77-1944	K-44	0092362	0092473

ITEMS with *: This data point is recorded only in the noted document(s).
ITEMS with **: This data point is recorded only in *Ordnance Department Administrative and Tactical Vehicles 1940-1945, Defense Aid.*

Table 4 - G-506 Production Records by Contract (cont.)

Continues →

QM Contract #	Ordnance Contract #	Prod Order Number	Bid	IFB Date	Proposal Date	Lettter Order Date	Contract Date	Award Date on WSC
517		T-14167						
517		T-14167						
517		T-10510						
517		T-14168						
517		T-10543						
517		T-21731						
33-019-4944		T-22635						
NOs95878								12.01.41
NOs2919								04.01.42
ONLY ON ATV & WSC*	W-374-ORD-6582	T-9134						

ITEMS with *: This data point is recorded only in the noted document(s).
ITEMS with **: This data point is recorded only in *Ordnance Department Administrative and Tactical Vehicles 1940-1945, Defense Aid.*

Completion Date on WSC	Year	Model	Plants Involved	Qty Ordered	Accepted per SRTAMA	Acceptance by year qty/yr	Vehicle Type	Start Reg No.	End Reg No.
			Atlanta, Baltimore, Flint, Norwood			538-1945	K-44	0098957	0099094
			Atlanta, Baltimore, Flint, Norwood				K-44	0099305	0099669
			Atlanta, Baltimore, Flint, Norwood	287	287	287-1944	K-44		
	1942	G-7103NE	Atlanta, Baltimore, Flint, Norwood	35	35	35-1945	cab and chassis wo/w	3506501	3507046
	1944	G-7103NE	Atlanta, Baltimore, Flint, Norwood	1562	200	200-1944	chassis for Corps of Engineers		
	1945	G-7107NJ	Atlanta, Baltimore, Flint, Norwood	104	104	104-1945	chassis for dump	3534366	3534469
			Hercules*		104	104-1945	Hercules steel contract for dump bodies		
04.01.42			Atlanta, Baltimore, Oakland, Tarrytown				not sure if this is G-506		
04.01.42			Baltimore				Not sure if this G-506		
	1942	G-7107NJ	Atlanta	10312	9450	9450-1944	cargo and 500 cab and chassis	3399524	3410335
			Atlanta		1362	1362-1944	cargo accepted as chassis		

ITEMS with *: This data point is recorded only in the noted document(s).
ITEMS with **: This data point is recorded only in *Ordnance Department Administrative and Tactical Vehicles 1940-1945, Defense Aid.*

Table 4 - G-506 Production Records by Contract (cont.)

Continues →

QM Contract #	Ordnance Contract #	Prod Order Number	Bid	IFB Date	Proposal Date	Lettter Order Date	Contract Date	Award Date on WSC
ONLY ON ATV & WSC*	W-374-ORD-6582	T-9134						
ONLY ON ATV & WSC*	W-374-ORD-6582	T-9134						
WSC only*	W-374-ORD-6072							05.01.43
WSC only*	W-374-ORD-5146							08.01.44
	535-AC-25544							05.01.42
	535-AC-32103							08.01.42
	W44-110-ENG-MSP-128							
	535-AC-31742							08.01.42

ITEMS with *: This data point is recorded only in the noted document(s).
ITEMS with **: This data point is recorded only in *Ordnance Department Administrative and Tactical Vehicles 1940-1945, Defense Aid.*

Completion Date on WSC	Year	Model	Plants Involved	Qty Ordered	Accepted per SRTAMA	Acceptance by year qty/yr	Vehicle Type	Start Reg No.	End Reg No.
	1942	G-7117NM	Atlanta	1944	1944	1944-1944	cargo w/w	3410336	3412279
	1942	G-7107NJ	Atlanta	500			chassis		
12.01.44			Detroit				Bomb Service Trucks		
11.01.44			Flint				M6 Bomb Service Truck Parts		
09.01.42			DAVEY				Davey Compressor field lighting units - perhaps J5		
11.01.42			DAVEY				Davey Compressor lighting units J5		
							3372900		
02.01.43							J-4 field lighting		

ITEMS with *: This data point is recorded only in the noted document(s).
ITEMS with **: This data point is recorded only in *Ordnance Department Administrative and Tactical Vehicles 1940-1945, Defense Aid.*

Table 5 - G-506 Production 1940-1944 per Chevrolet records (1945 records have not been located)

Plant code Number	8	9	2	14	6
	Atlanta 1940-44	Norwood 1940-44	Tarrytown 1940-42	Baltimore 1940-42	Oakland 1940-43
Cab & Chassis	1562	1718	2	18	
Panel		1430	335	994	56
Dump wo/w	686	2369	1979	1225	813
Dump w/w	203	479	2452	721	578
Cargo wo/w	35894	14623	6170	7475	9669
Cargo w/w	6753	3110	5620	2828	3371
LWB			50	166	81
Tractor		1713	598	297	274
COE		235		361	
Bomb service		5092		1311	
Crash		250		185	
Field Lighting					
Fire					
Earth borer		1119			
Telephone wo/w		67			
Telephone w/w		3661			
PRODUCTION					
TOTAL 1940	712	363	922	839	773
10.01.40	1	14	9	13	5
11.01.40	354	171	443	423	326
12.01.40	357	178	470	403	442

→ (continued) Table 5 - G-506 Production 1940-1944 per Chevrolet records

3	5		21		Plant code Number
St. Louis 1940-43	Kansas City 1940-42	Bloomfield 1942	Janesville 1940-41	Exp Lab 1941-42	
1					Cab & Chassis
663					Panel
1698	181		58		Dump wo/w
628	53		16	2	Dump w/w
3821	1685	264	237		Cargo wo/w
2295	1245		96	2	Cargo w/w
86					LWB
63					Tractor
				2	COE
1320				1	Bomb service
				1	Crash
					Field Lighting
					Fire
				2	Earth borer
					Telephone wo/w
				2	Telephone w/w
					PRODUCTION
1397	367		333		TOTAL 1940
42	11		7		10.01.40
661	124		196		11.01.40
694	232		130		12.01.40

Table 5 - G-506 Production 1940-1944 per Chevrolet records (1945 records have not been located)

Plant code Number	8	9	2	14	6
	Atlanta 1940-44	Norwood 1940-44	Tarrytown 1940-42	Baltimore 1940-42	Oakland 1940-43
TOTAL 1941	3233	1697	4777	4236	3294
01.01.41	374	183	306	522	468
02.01.41	340	159	122	415	309
03.01.41	34	39		137	
04.01.41	70	86	201	81	
05.01.41	290	57	226	282	131
06.01.41	305	92	246	376	377
07.01.41	306	111	358	444	392
08.01.41	391	56	245		218
09.01.41	350	125	756	439	433
10.01.41	257	204	879	493	326
11.01.41	257	287	791	601	322
12.01.41	259	298	647	446	318
TOTAL 1942	805	10958	11507	10504	8953
01.01.42	350	151	454	468	349
02.01.42	455	155	980	922	435
03.01.42		413	1301	1753	489
04.01.42		1049		1847	812
05.01.42		1311	8	2288	1059
06.01.42		1300	870	2064	1117
07.01.42		799	975	117	730
08.01.42		1008	1008	295	678
09.01.42		1415	2002	276	850
10.01.42		1043	1198	474	600
11.01.42		1148	1647		912
12.01.42		1166	1064		922

→ (continued) Table 5 - G-506 Production 1940-1944 per Chevrolet records

3	5		21		Plant code Number
St. Louis 1940-43	Kansas City 1940-42	Bloomfield 1942	Janesville 1940-41	Exp Lab 1941-42	
5615	2197		74	9	**TOTAL 1941**
739	240		74		**01.01.41**
518	189			3	**02.01.41**
					03.01.41
100	35			2	**04.01.41**
337	132				**05.01.41**
764	198				**06.01.41**
929	191			2	**07.01.41**
181	130				**08.01.41**
568	227			1	**09.01.41**
372	299				**10.01.41**
727	290			1	**11.01.41**
380	266				**12.01.41**
2243	600	264		2	**TOTAL 1942**
682	250			1	**01.01.42**
491	350				**02.01.42**
436					**03.01.42**
634					**04.01.42**
					05.01.42
					06.01.42
					07.01.42
		264			**08.01.42**
					09.01.42
					10.01.42
				2	**11.01.42**
					12.01.42

Table 5 - G-506 Production 1940-1944 per Chevrolet records (1945 records have not been located)

Plant code Number	8	9	2	14	6
	Atlanta 1940-44	Norwood 1940-44	Tarrytown 1940-42	Baltimore 1940-42	Oakland 1940-43
TOTAL 1943	9110	20297			1822
01.01.43		1614			950
02.01.43		1420			872
03.01.43		1983			
04.01.43		1979			
05.01.43		1905			
06.01.43	486	1949			
07.01.43	842	1643			
08.01.43	1156	1779			
09.01.43	1512	1591			
10.01.43	1801	1448			
11.01.43	1411	1326			
12.01.43	1902	1660			
TOTAL 1944	31476	2720			
01.01.44	2592	733			
02.01.44	2837	355			
03.01.44	2846	116			
04.01.44	2551	131			
05.01.44	2833	134			
06.01.44	2678	96			
07.01.44	2481	178			
08.01.44	2280	154			
09.01.44	3000	190			
10.01.44	2340	209			
11.01.44	2590	225			
12.01.44	2448	199			

→ **(continued) Table 5 - G-506 Production 1940-1944 per Chevrolet records**

3	5		21		Plant code Number
St. Louis 1940-43	Kansas City 1940-42	Bloomfield 1942	Janesville 1940-41	Exp Lab 1941-42	
1320					TOTAL 1943
50					01.01.43
430					02.01.43
415					03.01.43
434					04.01.43
11					05.01.43
					06.01.43
					07.01.43
					08.01.43
					09.01.43
					10.01.43
					11.01.43
					12.01.43
					TOTAL 1944
					01.01.44
					02.01.44
					03.01.44
					04.01.44
					05.01.44
					06.01.44
					07.01.44
					08.01.44
					09.01.44
					10.01.44
					11.01.44
					12.01.44

Table 5 - G-506 Production 1940-1944 per Chevrolet records (1945 records have not been located)

Plant code Number	8	9	2	14	6
	Atlanta 1940-44	Norwood 1940-44	Tarrytown 1940-42	Baltimore 1940-42	Oakland 1940-43
SHIPMENTS					
04.23.05	629	250	736	661	323
04.24.05	3266	1692	4911	4196	3588
04.25.05	855	10653	11387	10720	8826
04.26.05	8783	19869	172	4	2105
04.27.05	31565	3402			

Table 6 - G-506 Serial Number Prefix Decoder

	1941	1942	1943	1944	1945
January		01	13	25	37
February		02	14	26	38
March		03	15	27	39
April		04	16	28	40
May		05	17	29	41
June		06	18	30	42
July	07	19	31	43	
August	08	20	32	44	56
September	09	21	33	45	09
October	10	22	34	46	
November	11	23	35	47	
December	12	24	36	48	

The first number in any serial indicates the assembly plant at which the truck was produced.

The first two letters indicate the model series of the vehicle.

The number immediately following the model series indicates the month in which the truck was produced, according to the above table.

The last group of numbers indicates the sequence in which the unit was built. Each model series at each assembly plant is started with sequence number 1001.

Plant Number	Location
1	Flint, Mi
2	Tarrytown, NY
3	St. Louis, MO
5	Kansas City, KS
6	Oakland, CA
8	Atlanta, GA
9	Norwood, OH
14	Baltimore, MD
21	Janesville, WI

→ (continued) Table 5 - G-506 Production 1940-1944 per Chevrolet records

3	5		21		Plant code Number
St. Louis 1940-43	Kansas City 1940-42	Bloomfield 1942	Janesville 1940-41	Exp Lab 1941-42	
					SHIPMENTS
1062	304		204		04.23.05
5930	2159		203	9	04.24.05
2263	701	264		3	04.25.05
1320					04.26.05
					04.27.05

Table 7 - G-506 Reference Publications and Manuals

As is typical of equipment procured by the US military, an extensive library of documentation was produced detailing the use and maintenance of the G-506 series trucks.

At the time procurement of the vehicles began, the Quartermaster Corps was the governing authority (with the exception of the Bomb Service trucks), and the Quartermaster procedures were used.

That meant the Chevrolet was also contracted to produce manuals, a task that was handled by the Chevrolet Service and Mechanical Department. Today, surviving records tell us which manuals the company produced, and in many cases how many copies were printed. The first of these was a basic 4x4 manual, devoid of any "TM" number, titled "Maintenance Manual Chevrolet 1 ½-ton 4x4 truck (Conventional Models)." Interestingly the cover of the manual included a price for the 154-page manual – $1.00. Ultimately, most of these Chevrolet-produced manuals were assigned publication numbers by the Quartermaster Corps, and those numbers began with TM 10-.

When authority over procurement of transport vehicles was shifted to the Ordnance Department in 1942, the procedure for producing the manuals changed as well. The Ordnance format of the manuals were used, such that there were two series of manuals – those with three digit numbers, covering driver instructions and basic maintenance, and those with four digit numbers, covering third and fourth echelon maintenance and overhaul of all component units. Also, while the contract continued to require that the manufacturer write the manuscript, create the original artwork and provide photographs, this material was to be delivered to the Military Publications Section of the Ordnance Department, who then had the responsibility for printing the manuals. The Ordnance Produced manuals were assigned numbers in the TM 9- or ORD ranges.

While the bulk of the manuals produced fell under the auspices of the Quartermasters or Ordnance, there were some exceptions. Signal Corps equipment, notably the radio sets installed in K-51 vans and the earth boring equipment used on the K-44, were furnished with manuals in the TM 11- range, manuals for Engineer Equipment, including firefighting equipment, were in the TM 5- range, and equipment used by the Army Air Corps (and later, the Army Air Forces) were supported not with Technical Manuals (TMs), but rather Technical Orders (TOs).

Quartermaster Manuals – published by Chevrolet

Driver Instruction Booklet – Attached to the windshield with tape at delivery. Covers 1 ½-ton 4x4, 1 ½-ton 4x2, ¾-ton Trucks and Passenger Cars – 25,000 copies produced.

- Manual for regular production vehicles – 5,000 copies
- Supplement – 4x4 Cab Over Engine – 1,500 copies

- Supplement – Trucks equipped with Earth Boring and Pole Setting Equipment – 2,100 copies
- Air Field Lighting Trucks Manual – 5,500 copies
- Assembly Operations 4x4 Chevrolet Trucks Two-Unit Pack
- TM 10-1126 Parts Price List for Special Chevrolet 1 ½-ton Trucks four-wheel drive (4x4) built for the United States Army, September 1, 1941 – unknown print run
- TM 10-1126 Parts Price List for Special Chevrolet 1 ½-ton Trucks four-wheel drive (4x4) built for the United States Army, January 2, 1942 – unknown print run
- TM 10-1127 Maintenance Manual Chevrolet 1-1/2 Ton (LC), 4x4 Truck built for the United States Army, August 15, 1941 – 73,400 copies
- TM 10-1127 Camion Chevrolet 1-1/2 Tonne (LC), 4 Roues Motrices. Minstre de la Guerre. 19 May 1945
- TM 10-1130 Parts Price List, , Truck 1-1/2 Ton 4x4, Chevrolet (Model G-1723). War Dept. December 1941
- TM 10-1131 Maintenance Manual, Truck 1-1/2 Ton 4x4, Chevrolet (Model G-1723). War Dept. December 1, 1941 – 4,500 copies
- TM 10-1202 Parts Price List, Truck 1-1/2 Ton 4x4, Chevrolet (G-4112). War Dept. 1941
- TM 10-1203 Maintenance Manual, Truck 1-1/2 Ton (LC) 4x4, Chevrolet (Models G-7163 and 7173). War Dept. March 1942 – 7,500 copies
- TM 10-1203, Change #1, Maintenance Manual Chevrolet 11/2-ton 4x4 Truck, Supplement for Section 12 Special Equipment Telephone Maintenance Models G-4112 Series YP Contract Number W-398-QM-8778, June 15, 1941 – 1,500 copies
- TM 10-1203, Change #4 – 7,000 copies
- TM 10-1208 Parts Price List, Truck 1-1/2 Ton 4x4, Chevrolet (G-7103, G-7113). War Dept. 1941
- TM 10-1234, Parts Price List, Truck 1 1/2-ton, 4x4, Chevrolet G-7133 Maintenance 1942
- TM 10-1235 Truck 1 1/2-ton, 4x4, Chevrolet G-7133 Maintenance 1942 – 6,000 copies

- TM 10-1284, Parts Price List, Truck 1 1/2-ton, 4x4, Chevrolet G-7103 – G-7113), February 1, 1943
- TM 10-1414, Parts Price List, Truck 1 1/2-ton, 4x4, Chevrolet G-7163, G-7173), 1942
- TM 10-1430, Parts Price List, Chevrolet 1 ½-ton 4x4, G-4105, G-4113, G-4152, G-4162, G-4163, G-4165, G-4174
- TM 10-1431 Maintenance Manual Chevrolet 1 ½-ton 4x4 Truck Bid No 2-2810, June 28, 1942 – 15,000 copies
- TM 10-1438 Parts Price List for Special Chevrolet 1942 1 ½-ton (4x4) Trucks built for the United States Army, March 2, 1942 – unknown print run
- TM 10-1460 Parts Price List, Truck 1-1/2 Ton (LC) 4x4, Chevrolet (Model G-7116). War Dept. April 1942
- TM 10-1461 Maintenance Manual, Truck 1-1/2 Ton (LC) 4x4, Chevrolet (Model G-7105). War Dept. April 1942 – 22,000 copies
- TM 10-1475 Chevrolet Master Parts Price List 1941-1942 Passenger Cars and Trucks built for the United States Army, April 1, 1942 – unknown print run
- TM 10-1475 Chevrolet Master Parts Price List 1941-1942 Passenger Cars and Trucks built for the United States Army, October 1, 1942 – unknown print run
- TM 10-1557 Maintenance Manual, Truck 1-1/2 Ton 4x4, Chevrolet (Models G-7103, G-7105, G-7106, G-7107, G-7113, G-7116, G-7117, G-7163, G-7173). War Dept. July 1, 1942 – 64,500 copies
- TM 10-1558 Parts Price List, 1 ½ Ton (4x4) Cab-Over-Engine Truck with Platform and Stake Racks, Model G-7123, August 1, 1942
- TM 10-1559 Maintenance Manual, Truck 1 1/2-ton, 4x4, Chevrolet G-7123 August 1, 1942 – 1,000 copies

Ordnance Manuals – published by the Ordnance Department

- TM 9-765 Technical Manual, Bomb Service Truck M6 (Chevrolet), August 13, 1942
- TM 9-1764 Chevrolet M6 bomb service truck. 7 October 1941 – 2,100 copies
- TM 9-1765A Axles, Propeller Shafts, and Wheels for Bomb Service Truck M6 and 1½ ton, 4x4, Trucks (Chevrolet), including Telephone Earth Borer (K-44) December 1942
- TM 9-1765B Engine Power Train, Braking and Steering Systems for Bomb Service Truck M6 and 1½ ton, 4x4 Trucks (Chevrolet), February 1943
- TM 9-805 1 ½-ton 4x4 Truck (Chevrolet), 30 December 1943
- Winterization – Bomb Service – 2,100 copies
- Winterization – Other Variants – 4,500 copies
- Radio Suppression Equipment Manual – 4,500 copies
- ORD 7 SNL G-85 Organizational Maintenance Parts List, April 1948
- ORD 8 SNL G-85 Field & Base Maintenance Spare Parts and Equipment, January 1948
- ORD 9 G-85 Service Parts Catalog for Truck, Bomb Service, M6, dated October 14, 1945
- ORD 9 G-506 Service Parts Catalog for Truck 1 ½ Ton, 4x4 (Chevrolet) dated October 13, 1943
- ORD 9 SNL G-85 Vol IV and G-506 Ordnance Supply Catalog, Truck 1 ½-ton, 4x4, (SNL G-506), Truck, Bomb Service, M6 (SNL G-85 Vol IV), 25 June 1945

Signal Corps Manuals

- TM 11-280 Radio Set SCR-299-A Radio Set SCR-299-B Radio Set SCR-299-C Radio Set SCR-299-D, June 1943
- TM 11-280B Radio Set SCR-299-A and Radio Set SCR-299-B, War Dept, October 1942
- TM 11-364 Technical Manual, Truck K-44-B and Earth Borer Equipment HD, September 24, 1942

Engineer Manuals

- TM 5-315 Fire Protection by Troop Organizations in Theaters of Operation, March 23, 1944
- TM 5-316 Airplane Crash Fire Fighting, April 1944
- TM 5-687 Inspection and Preventative Maintenance for Fire-Protection Equipment and Appliances, January 1946

Engineer Tables of Organization and Equipment

- TO&E 5-337 May 12, 1943
- TO&E 5-500 Engineer Service Organization July 31, 1943

Army Air Forces Technical Orders

- TO 11-65-1 Type E-5 Aerial Gunnery Trainer handbook of Instructions with Parts Catalog, February 15, 1943
- TO 11-65-1 Type E-5 Aerial Gunnery Trainer handbook of Instructions with Parts Catalog, June 10, 1943
- TO 11-65-1 revision March 1, 1944
- TO 11-65-1 Type E-5 Aerial Gunnery Trainer handbook of Instructions with Parts Catalog, April 20, 1944
- AN 11-65-1 Type E-9 Aerial Gunnery Trainer handbook of Instructions with Parts Catalog
- TO 11-65-12 Handbook of Description Armament Training Devices, January 25, 1944
- TO 19-10-7 Handbook of Instructions with Parts Catalog Type J-5 Truck Field Lighting Unit, January 20, 1943
- TO 19-10-7 Handbook of Instructions with Parts Catalog Type J-5 Truck Field Lighting Unit, April 30, 1943
- TO 19-10-20 Handbook of Instructions with Parts Catalog Type J-4 Field Lighting Truck, January 25, 1944.

Table 8, G-506 Tool Sets

RESTRICTED

DEPARTMENT OF THE ARMY SUPPLY CATALOG **ORD 6 SNL J-16**

Section 19

Group J

Special Tool Set for Automotive Matériel (SNL Group G)

Section 19—Tool Sets, Field and Depot Maintenance

for

Truck, 1½-Ton, 4x4, Bomb Service, M6 (SNL G-85)

Truck, 1½-Ton, 4x4 (Chevrolet) (SNL G-506)

Supersedes that portion of ORD 6 SNL G-27, Section 1, 8 May 1947, and 8 November 1950, pertaining to the vehicles covered in this publication.

This catalog is correct to 17 August 1951

1. GENERAL. *a.* The tool sets listed herein are not issued for organizational maintenance. These tool sets are issued as required to ordnance maintenance companies and depots, ordnance offices of army and overseas command, and units of ordnance schools and replacement and unit training centers. This catalog supplements tables of organization and equipment, tables of basic allowances, and tables of allowances.

b. Explanation of symbols used in this catalog is included under NOTES.

c. Classification codes for Federal stock numbers listed herein are published in ORD 5-2, Alphabetical Index to Stock List of All Items.

d. Additional application of items listed in this publication, if any, are contained in the Department of the Army Supply Catalog ORD 5-5 series, Cross Reference List.

e. For prices of items contained herein, see the appropriate catalog of the ORD 5-3 series.

DEPARTMENT OF THE ARMY **SEPTEMBER 1951**

1

RESTRICTED

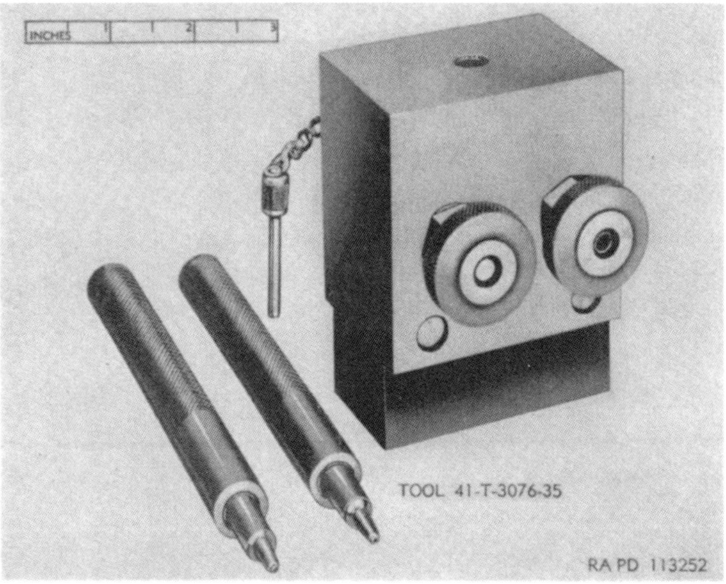

Figure 2. Field and depot maintenance tools.

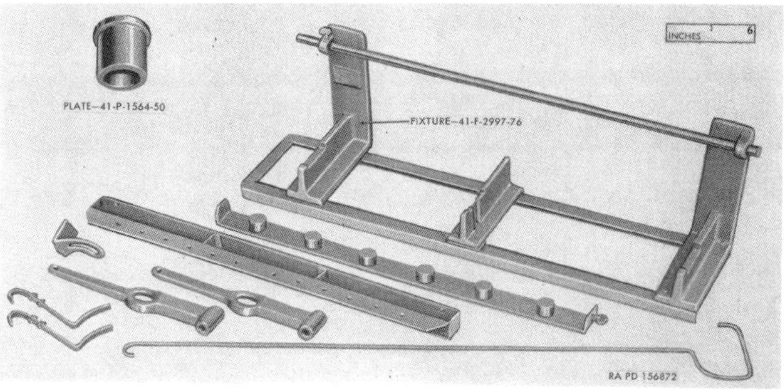

Figure 3. Field and depot maintenance tools.

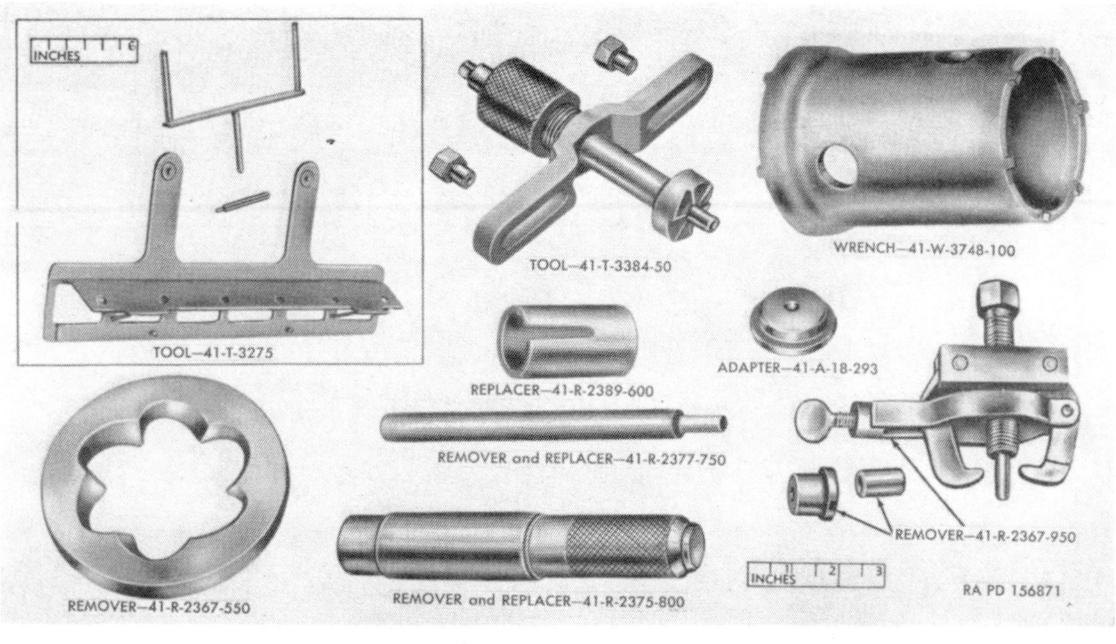

Figure 1. Field and depot maintenance tools.

Above and Below: *G-506 trucks in float livery passing in front of the Walter J. Bemb Buick-Pontiac, 6160 Cass Ave, Detroit, during a parade held during 1944.*